Palgrave Studies in Equity, Diversity, Inclusion, and Indigenization in Business

Series Editor
Eddy S. Ng, Queen's University, Kingston, ON, Canada

This series seeks to assemble scholarly research around equity, diversity, inclusion, and indigenization in the business world. The books in this series will explore contemporary diversity topics such as gender and corporate governance, sexuality and career development, social justice in the workplace, race and marketing, Indigenous inclusion, supplier diversity, mental health and neurodiversity at work, and more.

Through rigorous, disciplinary and interdisciplinary research and contributions from leading scholars, this series will provide an in-depth exploration of each topic and challenge scholars and practitioners to consider expanding their own awareness while addressing ways to make business (and society) more inclusive. The series speaks on behalf of anyone who has ever been affected–directly or indirectly–by discrimination or exclusion.

Marco Marabelli

AI, Ethics, and Discrimination in Business

The DEI Implications of Algorithmic Decision-Making

palgrave
macmillan

Marco Marabelli
Computer Information Systems
Department
Bentley University
Waltham, MA, USA

ISSN 2731-7307 ISSN 2731-7315 (electronic)
Palgrave Studies in Equity, Diversity, Inclusion, and Indigenization in Business
ISBN 978-3-031-53918-3 ISBN 978-3-031-53919-0 (eBook)
https://doi.org/10.1007/978-3-031-53919-0

This Palgrave Macmillan imprint is published by the registered company Springer Nature Switzerland AG
The registered company address is: Gewerbestrasse 11, 6330 Cham, Switzerland

Preface

This book documents inequalities and social injustices that stem from using algorithmic-led automated systems such as artificial intelligence (AI) in irresponsible ways. These widespread and ubiquitous systems permeate and monitor the minutiae of our everyday life. To do so, they rely on sensors (wearable devices such as smartwatches that individuals wear to improve their health) but also monitoring devices such as surveillance cameras in the workplace and on the street, recommendation systems on websites, data-extraction software in virtual reality platforms (which themselves involve sensors as users wear headsets and even bodysuits to access these platforms), and more. Social injustices can lead to ethical and diversity, equity, and inclusion (DEI) issues, which I unpack throughout the book with examples from the society in which we live.

To explain the ways that algorithms and AI can lead to social injustices, I take a historical perspective and refer often to laws, with a particular emphasis on the American system. I focus on the US[1] because, more so than most countries, it has a contradictory past: even after it ostensibly made slavery illegal, people in power have continued to perpetuate structural/institutional[2] racism for most of the past century, and governments

[1] Throughout the book, I use America and US interchangeably.

[2] As most literature agrees that the terms structural and institutional (racism) can be used interchangeably (cf. Bridges, 2019), in the reminder of the book I will only use the term structural.

and institutions have done little to terminate segregation practices. Structural racism falls under the umbrella of critical race theory (CRT), which I embrace. CRT is a framework and a theory that suggests that discrimination occurs because current laws and accepted social practices somehow enable discrimination (structural racism) or because certain actors and institutions ignore or bypass anti-racist laws. All this stems from the past, where (US) institutions were built by (and designed for) Whites. In this regard, I argue that today's algorithms still enable and perpetuate racial discrimination. Companies, which are revenue-driven, exploit the lack of regulations concerning AI systems, thereby reproducing racism using historical and current data. Historical data is problematic because they do not reflect individual rights in the current time, and current data is problematic because they (might) reflect structural racism (cf. CRT). Along with racial discriminations, I discuss other issues concerning privacy (which, by the way, can lead to discrimination), gender discriminations, and other forms of exclusions affecting people at the margin, which are often magnified by automated systems.

To this end, I also criticize the laissez faire approach that America increasingly embraced from the 1940s—an approach that followed many principles from writer and philosopher Alice O'Connor (better known as Ayn Rand). This laissez faire approach in the mid-1900s, which contrasted with (and partially suppressed) President Franklins D. Roosevelt's social systems and the New Deal, heavily reduced government intervention in people's work and social life. Backed up by the Reagan Administration, in the 1980s, the laissez faire approach still prevented the US Governments from effectively regulating businesses such as high-tech companies. As a result, in the US (a market with few regulations), many actors now use AI and algorithms in ways that perpetuate historical injustices (e.g., recommendation systems specifically targeting people at the margin in the name of profit, capitalism, and a free market). I don't think that, in the era of generative AI, Milton Friedman's free-market theory still stands. Furthermore, the prominent role that America (still) has in creating and developing cutting-edge technologies affects the world despite efforts to "contain" the use AI in widespread and irresponsible ways (e.g., in the European Union (EU)).

With that said, I explore why AI systems so readily generate social injustices. One issue that I see as prominent concerns AI's digital (and *binary*) nature, which can conflict our analog life. Binary systems relying on classification processes can mirror reality only to a certain extent (i.e., people are Black, White, Brown, male, female, able, differently able, rich, poor, executives, managers, task workers, unemployed). Such classifications lack nuance and do not tell the stories behind people and their social conditions and struggles. As Bowker and Star (2000) explain, human beings take their own classification processes for granted, which leads to what they call infrastructure "naturalization." Yet, we should not take for granted how algorithms and AI classify phenomena, as it currently happens with automated systems used to assess employee performance. Algorithms used in the workplace for performance management purposes do not account for "soft" indicators (e.g., they would typically not know that a salesperson traveled less in recent months due to pregnancy since they would not have access to such data). But allowing automated systems to capture and process personal information would violate individuals' privacy, which leads to a tradeoff between privacy and discrimination. Less privacy might lead to more informed algorithms and fairer assessments.

The same happens with social media and discriminatory ads that offer well-paid jobs to White men if compared with jobs offered to women and Black people. Algorithms have no clue about race and gender. They follow patterns and discover that some individuals have, for example, purchased expensive tickets for the last Opera concert in town. Based on that data, they might assume that these individuals have disposable income and a good job and, thus, that they might be interested in well-paid jobs. However, online behaviors on social media can serve as proxies for race and gender. Algorithms do not know anything about race and gender, but they can make racist and discriminating inferences. Again, feeding algorithms with details about users' race and gender details would mean one needed to collect personal information, which leads to privacy issues. In this book, I discuss the tradeoff between privacy and discrimination, the way that systems harvest, and algorithms classify information, and the role that governments and institutions play in addressing (or not) such issues.

On social media, algorithms manage online information and show people what they care about the most without regard for importance (Facebook's News Feed perhaps represents the first algorithmic-based system responsible for creating a filter bubble[3]). For instance, how many people know who Abiy Ahmed or Vladimir Putin are? Ahmed, an

[3] https://www.technologyreview.com/2023/11/14/1083336/how-facebook-went-all-in-on-ai/.

Ethiopian politician, has served as the prime minister for Ethiopia since 2018 and led the Prosperity Party since 2019. Meanwhile, Putin, a former KGB officer, currently serves as the president for Russia. Both started a war in recent times. Most sources agree that, in 2022, nearly 10,000 people died in the Russia-Ukraine War.[4] In the same year, an estimated 500,000 people died in the Tigray War in Ethiopia according to the United States Ambassador to the United Nations Linda-Thomas Greenfield.[5] It is with no doubt that the Russia-Ukraine War has drawn more attention due to the associated geopolitical consequences. Still, while one can find some news about the Ethiopia civil warin the printed press, it rarely appears in social media feeds (i.e., a primary news source for many people these days). Among other reasons, I still get *The New York Times* physically delivered to my door every morning because printed press does not get ordered in an algorithmic way (as often happens with online news outlets based on the articles that I've read in the past month or so). While I do not argue that news about the Ethiopian civil war and the Russia-Ukraine War concern algorithms exclusively, automated systems very much drive demand for news content.

In 2006, the US Patent Office received a request to file a patent whose abstracts read as follows[6]:

> *A system and method provides dynamically selected media content to someone using an electronic device in a social network environment. Items of media content are selected for the user based on his or her relationships with one or more other users. The user's relationships with other users are reflected in the selected media content and its format. An order is assigned to the items of media content, for example, based on their anticipated importance to the user, and the items of media content are displayed to the user in the assigned order. The user may change the order of the items of media content. The user's interactions with media content available in the social network environment are monitored, and those interactions are used to select additional items of media content for the user (US8171128B2).*

[4] https://www.ohchr.org/en/news/2023/07/ukraine-civilian-casualties-24-february-2022-30-june-2023.

[5] https://english.elpais.com/international/2023-01-27/ethiopias-forgotten-war-is-the-deadliest-of-the-21st-century-with-around-600000-civilian-deaths.html (here, the numbers vary by source and span from 300,000 to 800,000, so the US Ambassador's estimate is probably rather accurate).

[6] https://patents.google.com/patent/US8171128B2/en.

Mark Zuckerberg, Andrew Bosworth, Chris Cox, Ruchi Sanghvi, and Matt Cahill filed the patent on behalf of Facebook. In my opinion, this patent represents one of the most revolutionary "inventions" (the definition used when filing patents) in the past few decades (probably only second to the Internet). An AI system called EdgeRank powered this system to show Internet users feeds based on "relevance" rather than chronological order (the latter already happened in other social platforms such as MySpace in 2003). Shortly after its creation, a machine learning (ML) system powered EdgeRank, in 2011. While AI and ML (a branch of AI that operates without specific instructions from programmers) have been around since the 1950s (IBM computer scientist Arthur Lee Samuel first coined the term "machine learning" in 1959), only with the availability of massive amount of user-generated data (social software) could companies such as Facebook (now Meta) develop algorithms that literally manage our everyday life today. The News Feed patent and the following EdgeRank AI system pioneered what we now call classification algorithms that go beyond prioritizing content and news on social media, and have now spread in many other settings of our lives. These (and many other) algorithms constantly act in discriminatory ways based on real-time and historical (often skewed) data in addition to posing privacy concerns.

Not surprisingly, algorithm- and AI-powered discrimination and privacy issues affect people unevenly. Discrimination applies to underrepresented populations almost by default. With respect to privacy, in many circumstances, people at the margin give up their personal data. Examples include people out on parole who need to wear anklets, drivers who want to lower their car insurance premium by installing on-board devices such as the one that Progressive Insurance[7] (among other companies) have commercialized, patients who rely on public healthcare such as pregnant women (cf. Bridges, 2011), and so on.

One pressing, related issue that I investigate in this book concerns how all actors collect this data and how discrimination operates. While, in the above examples, people typically recognize they have given up their data, other settings lack clarity on the issue. For instance, we do not know how much data our banks collect on our spending behavior (they initially collect it to prevent fraud but it often leads to decisions on our credit line). In the past, the US Government used data

[7] https://www.progressive.com/answers/telematics-devices-car-insurance/.

that banks harvest to discriminate against Black communities, such as when the US Federal Housing Administration recommended that lenders prevented Black people from taking loans to purchase houses in White neighborhoods in the 1940s and 1950s. Nowadays, lenders perpetuate this discrimination when granting mortgages based on algorithms that account for an applicant's zip code often a proxy to race and socioeconomic status. Almost a century ago, discrimination was a top-down process, intentionally initiated by the federal government. Nowadays, the government is complacent (cf. laissez faire), enabling structural racism.

As powerful institutions such as banks and large companies use algorithms and so-called people analytics techniques to control people (e.g., Amazon uses cameras in their warehouses to monitor employees and reward or punish them), some people have learned to game the algorithms at hand in response. For instance, Uber drivers gather in forums to share their frustrations using the Uber app and share workarounds, salespeople identify key performance indicators (KPIs) that managers look into to reward them (or not), and borrowers figure what financial indicators loan managers (or algorithms on their behalf) use and act accordingly to look good and obtain a loan.

One main focus of this book concerns the ethical and DEI issues around how institutions (laws and regulations) can (or cannot) limit the power of companies that continuously mine individuals' personal data and infer algorithmic-based conclusions that often penalize minorities. I can understand that some governments may find it hard to formulate (and pass) specific laws regulating AI because lawmakers often lack "hard evidence" on automated systems' negative consequences. For instance, while many people may know about the negative consequences that can arise from becoming addicted to social media (due to anecdotal evidence), most people cannot access the supporting data because social media companies such as Meta own it.

What we do know comes from whistleblowers such as Frances Haugen, former Meta employee, who exposed social media's negative effects. In 2021, Ms. Haugen reported on Meta's internal research that showed the company knew very well how its platforms (e.g., Instagram) harmed teenagers with eating disorders and often exacerbated their health issues by sending them to social media pages that suggested strict diets.[8] Similarly, we do not know exactly whether (or how) Amazon abuses its dominant position by setting interlocking anticompetitive and unfair strategies to illegally maintain its monopoly power. However, based on complaints from many retailers, the Federal Trade Commission (FTC) led a lawsuit[9] recently on Amazon's alleged dominant position in the retail market. Some governments have taken strong action to limit high-tech companies' ability to control people's data and lives and jeopardize their privacy. Notable examples include the European Union (e.g., with its General Data Protection Regulation (GDPR)[10] and AI Act[11]) and China (e.g., with its personal information protection law (PIPL)[12]). These efforts made me wonder why the US has lagged behind so much on AI-related regulations, until I came across a 2021 article that reported Eric Emerson Schmidt's (former Google CEO) asking: Why don't we wait until something bad happens and then we can figure out how to regulate it—otherwise, you're going to slow everybody down. Trust me, China is not busy stopping things because of regulation.[13] Similarly, during a speech in 2022, Tesla CEO Elon Musk noted that: "In the US things are legal by default, and in Europe they're illegal by default.[14]"

Relentlessly focusing on monetizing emerging innovations, especially in the short term, poses particular dangers if we think about generative AI. Italy paused ChatGPT for an entire month in April 2023 because OpenAI allegedly collected users' data without their consent, and failing to prevent underage users from accessing inappropriate material.[15] In the US, the onus to protecting privacy falls on companies and individuals. For

[8] https://www.wsj.com/articles/facebook-knows-instagram-is-toxic-for-teen-girls-company-documents-show-11631620739.

[9] https://www.ftc.gov/legal-library/browse/cases-proceedings/1910129-1910130-amazoncom-inc-amazon-ecommerce.

[10] https://gdpr.eu.

[11] https://www.europarl.europa.eu/news/en/headlines/society/20230601STO93804/eu-ai-act-first-regulation-on-artificial-intelligence.

[12] http://en.npc.gov.cn.cdurl.cn/2021-12/29/c_694559.htm.

[13] https://www.youtube.com/watch?v=CmBpsw1ORQ0.

[14] https://slate.com/technology/2023/02/tesla-recall-full-self-driving-nhtsa-musk-regulation.html.

[15] https://www.theverge.com/2023/4/28/23702883/chatgpt-italy-ban-lifted-gpdp-data-protection-age-verification.

instance, in May 2023, Apple restricted employees from using ChatGPT due to fears that they would leak industry secrets because employee prompts (way to ask questions to generative AI) could leak research that the company conducts.[16]

However, organizations will not just use generative AI to harvest Web knowledge (such as how Google has used it with very questionable results[17]). Generative AI will assist, if not power other trends such as virtualized interactions (at work or school) via 3D immersive platforms such as the metaverse (e.g., with AI avatars that chat with "real" people in virtual environments). These and others uses of AI pose ethical and DEI issues. For example, to train generative AI to directly interact with people, companies typically use data collected (and often brutally scraped) from the Internet. This user-generated data can feature significant bias and dangerous (hate speech) and nonfactual (fake news) content. Even when one trains AI systems with sources such as Wikipedia, which many view as the gold standard for truthful information, we need to remember that Wikipedia content (at least in its English version, which companies mostly commonly used to train AI) comes from contributors mainly from Western countries and that females amount to only 8-15% of all contributors. One needs to consider the consequences from training AI-powered avatars with potentially skewed data and that interact, in real time, with "real" people (also using avatars) in immersive platforms. Racist AI in the metaverse could have much more harmful impacts than AI that generates racist content on social media due to virtual platforms' immersive nature. We also need to consider the fact that AI often makes up facts (so-called "hallucinations") and that the AI supply chain often involves training that partially relies on people from Global South countries who lack qualifications for the work but, more importantly, often receive poor pay and face exploitation—an ethical issue.

With respect to immersive platforms, accessibility can raise important DEI concerns. In general, since one needs expensive equipment (e.g., headsets and bodysuits not to mention the need for sufficient bandwidth) to access immersive platforms, only certain people will be able to benefit from them. Thus, not everyone would be able to benefit from immersive

[16] https://www.wsj.com/articles/apple-restricts-use-of-chatgpt-joining-other-companies-wary-of-leaks-d44d7d34.

[17] https://hbr.org/2023/02/generative-ai-wont-revolutionize-search-yet.

platforms in, for example, a healthcare or education setting, (where the gap between community colleges and Ivy League Schools would widen). Similarly, virtual reality in the workplace could promote remote work with interactions that resemble in-person interactions with the caveat that task workers such as bus drivers, grocery cashiers, home cleaners, and so on would not benefit (and, often, such jobs serve as proxies for race and gender, which adds to the potential for virtual platforms to discriminate against people).

Technologies reflect our society and who we are. Thus, algorithmic-powered AI systems unsurprisingly serve the powerful and leave behind people at the margin. One important issue that I tackle throughout the book concerns whether (and when) companies purposely create AI to discriminate or whether the problem rests on the fact that the companies developing AI care so much about revenue that they become blind to social injustice. For instance, some companies use AI-powered hiring software to avoid the need to manually scan applications or hold Zoom interviews. However, as a result, the software can reject "unconventional" (e.g., differently able, transgender, etc.) candidates even when they have the necessary qualifications for jobs that they apply for. In this case, companies may not aim to purposely discriminate. Yet, their focus on efficiency (and their negligence) results in it.

People who advocate for idea that some companies do intentionally discriminate—such as Safiya Umoja Noble, the David O. Sears Presidential Endowed Chair of Social Sciences and Professor of Gender Studies, African American Studies, and Information Studies at the University of California, Los Angeles (UCLA)—have provided examples of companies that have purposely designed racist AI systems (or companies that knew that they used racist systems—do the two differ?). Algorithms that support predictive policing and target poor and Black neighborhoods in large US cities such as Los Angeles and Chicago embed racial bias in their design. *Ghetto* neighborhoods in the US stem from past injustices from almost a century ago when redlining (unlawful) policies prevented Black Americans from purchasing homes in White neighborhoods due to the so-called "detrimental influence of the concentration of Negros" in certain neighborhoods.[18] People such as Yanni Alexander Loukissas, Founding

[18] https://controller.phila.gov/philadelphia-audits/mapping-the-legacy-of-structural-racism-in-philadelphia.

Executive Director of the Interdisciplinary Media Arts Center and Associate Professor of Digital Media, Georgia Institute of Technology, have argued that algorithmic-based real estate websites such as Zillow[19] reflect past injustices and noted that: "The entire real estate market that Zillow aims to optimize is a system by which society segregates itself to preserve social hierarchies, such as class and race.[20]"

In the book, I jointly examine how actors have both intentionally designed and used racist AI systems because one cannot readily distinguish between them without reading into the individuals who designed the algorithms that underpin them. In the past, with physical artifacts such as transportation infrastructures, one could more easily spot (often successful) attempts to discriminate people based on their race. A famous (and infamous) example concerns the "Moses bridges" in Long Island in the 1950s. Robert Moses, Long Island State Park Commissioner at the time, purposely designed the bridges to have a low clearance so that cars could drive under, but buses could not. As a result, rich people with cars could access Long Island areas nice parks and playgrounds for kids, while poor people who did not own cars and had to use public transport (often Black Americans) could not. Nowadays, one cannot easily spot racist-by-design algorithms. But one can readily find algorithms that focus on maintaining the status quo (White, male predominance), I document in the book. While still racist, such algorithms discriminate in a more subtle way that often involves structural racism, waives individuals' *intentionality* (such as the Moses Bridges), and showcases the consequences that result from laws or even common practices that do not account for disadvantaged people. A lack of intentionality connotes structural racism. An exemplary quote from Hamilton and Ture (1967) almost explains itself:

> When white terrorists bomb a black church and kills five black children, that is an act of individual racism widely deplored by most segments of society, but when in that same city—Birmingham, Alabama—five hundred black babies die each year because of the lack of proper food, shelter and medical facilities ... that is a function of institutional racism.[21]

[19] https://www.zillow.com.

[20] https://dl.acm.org/doi/pdf/10.1145/3529958.

[21] Retrieved from Bridges (2019).

Current examples of structural racism involve public schools in the US, which real estate taxes predominantly fund. In poorer neighborhoods with low real estate values, public schools receive less money than public schools in more affluent neighborhoods. Schools in less affluent neighborhoods, in turn, have less and/or poorer infrastructure, books, and equipment. Poor neighborhoods often serve as a proxy for race—Black Americans or other minorities often prevalently inhabit them. Black children are more likely to attend these schools, benefit from fewer and/or poorer resources, not perform well in standard tests that they need to access good colleges, and access well-paid jobs. Among other reasons, this situation explains why, until the 2022 Students for Fair Admissions, Inc. v. President and Fellows of Harvard College[22] US Supreme Court ruling, colleges in the US often acted in a race-conscious and affirmative manner when considering admissions. The COVID-19 pandemic showed that, when in March 2020 schooling worldwide went online, students who attended schools in affluent neighborhoods in the US could attend classes remotely because schools could provide laptops and learning platforms and train teachers to use them. Others students attending schools in poor neighborhoods often lacked similar opportunities as I illustrated in a recent study (Marabelli et al., 2021). Thus, one can see how a failure to change segregation practices from over a century ago continues to discriminate against certain individuals in society.

In conclusion, structural racism contributes to maintaining the status quo, and algorithms that adopt a design that reproduce past injustices, are racist by definition. Sticking with the schooling example above, as long as we use algorithms fed with variables such as scholastic aptitude tests (SATs) to make decisions about who to admit to colleges, we will perpetuate social injustice. The algorithm that the Zillow website that I mention above uses, exemplifies an algorithm that focuses on reproducing segregation. With the advent of ***datafication***, the shift from physical to digital artifacts as the center of our social life has not prevented people in power from taking advantage of their dominant position. Moses' bridges have now become AI-powered algorithms that prevent some people from having a fair shot at getting good education, buying a house, and being successful. In this book, I unveil the subtle and often invisible effects that algorithms and AI have on today's society while proposing solutions

[22] https://www.supremecourt.gov/opinions/22pdf/20-1199_hgdj.pdf.

that mainly rely on the idea of abandoning free-market principles and advocating for government intervention to regulate the ways in which individuals and companies design, develop, and use emerging algorithmic-centric technologies.

In Chapter 1, I focus on classification issues—the main reason why automated systems such as AI and algorithms reproduce past discrimination. In the apartheid era in South Africa, the government used to systematically classify people based on race. Today, banks continue to classify people based on their race (e.g., via their zip code, income, etc.). In the chapter, I provide examples from the UK higher education system, healthcare in the US and UK, AI-powered hiring software, the US juridical system, and access to credit. I stress the fact that classification, race, and gender constitute socially constructed phenomena. I conclude that it is possible to mitigate automated discriminations by focusing on team diversity, ethical design of algorithms, and more.

In Chapter 2, I discuss data-extraction practices such as sensors and Web scraping and question practices performed by data extractors (power companies that use harvested data to make inferences on large populations). Along with privacy issues, extracting data via sensors equates with surveillance, especially in the workplace. To this end, I discuss the people analytics literature and workplace discrimination. I also discuss India's digital identity initiative (an anti-poverty initiative that nevertheless controls people and, in some cases, exacerbates power discrepancy between everyday people and the government) and the poorly regulated data broker market in the US. I conclude that data has become currency that predominantly governments and powerful companies own.

In Chapter 3, I focus on data training; pinpoint the difference between manual training, which exploits the workforce in the Global South, and automated training, which relies on Internet data (a "place" with inherent biases) and Web scraping; and raise major privacy and discrimination concerns with these practices. I also consider the environmental issues associated with training complex ML models, bitcoin transactions, and so on. I conclude by noting the DEI issues that often emerge with data training (e.g., exploited underage workers in remote countries).

In Chapter 4, I point to the need to account for surveillance practices when monitoring people with sensors and other technologies. I analyze surveillance in professional bureaucracies and discuss a healthcare organization that implemented a people analytics initiative to improve its clinicians' and staff members' internal processes.

I contrast such surveillance with that happening to gig economy workers, who often face exploitation while living at the margin. I also discuss surveillance via algorithms in social media settings and how blind algorithms become racist due to their design. I conclude with thoughts concerning algorithms' transparency and whether power asymmetry between the surveillers and the surveilled is ethical and socially acceptable.

In Chapter 5, I draw from America's political history to explain how structural racism survives nowadays through algorithms. I also compare current US laws with laws in the EU and China. I claim that the AI industry cannot regulate itself in a free-market environment as Nobel Laureate Milton Friedman envisioned, and I advocate for strong, top-down government policies to regulate emerging AIs such as generative AI. I conclude that governments today need to oversee innovations, given the power that high-tech companies hold. Notwithstanding ethical and DEI issues, companies will recreate social injustices due to their revenue-led nature if they do not face discipline.

In Chapter 6, I examine emerging ways to use AI (and generative AI in particular) with respect to immersive platforms such as the metaverse. After the COVID-19 pandemic, 3D virtual environments have seen increasing use. But what DEI implications will these platforms have? Who will be left behind? What will companies do with data harvested in the metaverse? We need to ask and address these important questions before it is too late. Other disruptive innovations such as social media surprised everyone and have not faced significant regulations for years. With immersive platforms such as the metaverse (not here yet), we have a chance to understand how they work before the companies deploy them at a large scale and, thus, to prevent ethical and DEI issues from spreading. I conclude the chapter with recommendations on how to design and deploy immersive platforms responsibly.

I conclude this preface with some notes for the reader. First, even if I hope that the reader will go through this book from head to toe, because each chapter talks about a specific issue, some might approach this book selectively, i.e., by reading only some chapters. For this reason, in each chapter I spell out main acronyms and make sure that references to other chapters are highlighted. Second, the book has plenty of references and footnotes. While I use a typical academic style (in-text reference naming authors) for academic work, I use footnotes for press articles and other online articles of various type such as references to government websites, US Supreme Court rulings, etc. Third, appendixes are provided that group all US Supreme Court rulings along with other US and international laws and regulations that I mention in the book.

Waltham, USA Marco Marabelli

References

Bowker, G. C., & Star, S. L. (2000). *Sorting Things Out: Classification and Its Consequences*. MIT Press.

Bridges, K. (2011). *Reproducing Race: An Ethnography of Pregnancy as a Site of Racialization*. University of California Press.

Bridges, K. (2019). *Critical Race Theory: A Primer*. Foundation Press.

Hamilton, C. V., & Ture, K. (1967). *Black Power: Politics of Liberation in America*. Vintage Books, Random House.

Marabelli, M., Vaast, E., & Li, L. (2021). Preventing Digital Scars of Covid-19. *European Journal of Information Systems, 30*(2), 176–192.

Acknowledgements

Several individuals and institutions supported me in writing and finalizing this book, both financially and by providing invaluable feedback.

Bentley University gave me the time to write (time being the most precious resource, when dealing with a single-authored book) and supported my writing with a generous summer grant in 2023. In addition, my institution is very sensitive to DEI issues and constantly offers seminars and workshops to faculty—I attended several and found them very informative and helpful. I am also very lucky to be able to teach a graduate class on data analytics and ethical/DEI implications, where I challenge students on delicate topics concerning data, algorithms, and discriminations. I've learned a lot from class discussions, student comments, and written essays. A special thank goes to my students over the past few years! And to my department chair, Professor Monica Garfield, who made all this happen.

An early version this book was presented at HEC Montreal, where the information systems department faculty and Ph.D. students provided meaningful feedback. I would like to thank, in particular, Benoit Aubert, Camille Grange, Stefan Tams, and Gregory Vial. Also, thanks Camille for taking care of me for the whole day!! During my visit at HEC Montreal, I met Pamela Lirio, associate professor at the University of Montreal, with whom I started a collaboration leading to joint work including the organization of a workshop on the DEI challenges and opportunities of the metaverse, during the 2023 the Academy of Management meeting, which informed several parts of my book. Pamela too was key with her feedback on my view of AI and DEI, specifically with respect to gender issues.

An early version of selected chapters of the book were presented at the 2023 Academy of Management Annual meeting, in Boston, to the Bentley Ph.D. students (2023 seminar), and the 2023 Bentley Research showcase and posters presentation.

Other people provided feedback on various chapters (or portions of them) of my book. These include Kathy Chudoba, Dhaval Dave, Lydia Li, Mareike Möhlmann, Sue Newell, Sam Zaza, and many others that I apologize in advance for forgetting to mention.

A special thank you goes to some wonderful individuals who made themselves available to be interviewed for my book. These include (in alphabetic order) Donna Maria Blancero, Khiara Bridges, Srravya Chandhiramowuli, Noah Giansiracusa, Beryl Lipton, and Silvia Masiero.

Praise for *AI, Ethics, and Discrimination in Business*

"The ascent of AI affords a unique opportunity to assess and redress past wrongs. Marco Marabelli's sweeping overview of contemporary issues can help us prevent today's challenges from following us into the future."
—M. Lynne Markus, *Professor Emerita of Computer Information Systems, Bentley University*

"'*AI, Ethics, and Discrimination in Business: The DEI Implications of Algorithmic Decision-Making*' is an eye-opening volume on the human and social consequences of algorithmic injustice. Marabelli's historical perspective unpacks data violence in ways that range from data extractors to AI training, social media and workplace surveillance, at the same time casting a powerful light on the resistance practices of the surveilled. The book's lens of structural racism, substantiated in the tools of AI, provides a perspective that positions Marabelli as a leading voice in today's debate on data injustices."
—Silvia Masiero, *Associate Professor of Information Systems, University of Oslo and Editor in Chief, Information Technology for Development*

"In the rapidly evolving landscape of technology and business, the convergence of artificial intelligence (AI) and ethical considerations has emerged as a critical area of exploration. In '*AI, Ethics, and Discrimination in Business: The DEI Implications of Algorithmic Decision-Making*' Dr. Marco Marabelli exposes the historical roots of algorithmic racism and questions the lack of U.S. regulations regarding AI systems. Through a blend of academic research and accessible narratives, Dr. Marabelli identifies the pernicious effects of relying on algorithmic and AI systems that are fed biased historical data, rely on rough categorizations, and require invasions of privacy to even begin to develop a nuanced understanding of society. This work is indispensable for scholars, practitioners, policymakers, and anyone committed to fostering an equitable technological future. As Dr. Marabelli insightfully reminds us, '*Technologies reflect our society and who we are. Thus, algorithmic-powered AI systems unsurprisingly serve the powerful and leave behind people at the margin.*'"

—Melissa Mazmanian, *Professor and Chair, Department of Informatics in the School of Information and Computer Sciences, and Professor of Organization and Management, Paul Merage School of Management (joint), University of California, Irvine*

"The book offers a much needed, detailed exposition of the many negative impacts of AI and AI algorithms on social justice. Using examples, extensive references from research projects and the popular press, and his own research, Marco Marabelli persuasively demonstrates how AI and AI algorithms based on real-time and skewed historical data can perpetuate deeply embedded systemic societal injustice, including racial discrimination. He calls for remedies in the form of ethically-driven algorithmic design to mitigate all forms of discrimination, responsibly-deployed immersive platforms, and strong top-down government policies to regulate these and emerging technologies such as generative AI. Marabelli is to be applauded for skillfully championing this controversial issue."

—Carol Saunders, *Professor Emerita of Management, University of Central Florida*

"AI remains largely invisible while it touches every aspect of our lives, raising crucial questions about its potential impacts, particularly on discrimination. This book does a great job of unpacking these issues in a business context, explaining the technology in clear terms, and providing compelling examples of the promise and perils of these increasingly influential systems. I can't think of a more important topic for our times."

—Susan Winter, *Associate Dean for Research, College of Information Studies, University of Maryland*

Contents

List of Figures

List of Tables

CHAPTER 1

Zeros and Ones: Striving to Classify

While the world in which we live is analogic, what computers tend to mimic (e.g., AI) is digital/binary. From an ethical standpoint, it is acceptable (and often helpful) to generate digital files of various nature (documents, photos, audio/video) that reflect reality. However, recent AI applications such as facial recognition systems, emotion detection, and so on illustrate the limits (and challenges) of relying on binary systems to scrutinize an analogic world. In this chapter, I argue that the binary nature of computers does not do justice to the analogic world in which we live.

Classification and Marginalization

Classifying is idiosyncratic to human nature. In Europe, ancient Greeks used to classify stars, planets, and atmospheric phenomena such as the intensity of precipitations. In the Americas, the Aztecs, Maya, and Incas mastered astronomy and provided classifications of natural phenomena that are still difficult to achieve with modern instruments. Ancient Egyptian societal classifications included pharaohs, government officials, nobles and priests, scribes and soldiers, artisan and merchants, and slaves. In the United States (US), societal classification (slaves, masters) began with the colonization of parts of North America in the 1500s and 1600s. As Isabel

M. Marabelli, *AI, Ethics, and Discrimination in Business*, Palgrave Studies in Equity, Diversity, Inclusion, and Indigenization in Business,
https://doi.org/10.1007/978-3-031-53919-0_1

Wilkerson argues in her recent book *Caste*, "before there was a United States of America, there was the caste system" (Wilkerson, 2023, p. 41).

One prominent classification theory (or a critique of classification) in sociology that has propagated in business and computer science disciplines comes from Bowker and Star (2000) in their seminal book *Sorting Things Out*. They provide convincing arguments in support of humankind's need to classify, yet they also offer examples of how classification has historically led to, for example, segregation practices in South Africa. Overall, classification using binary systems provides an opportunity to make efficient assessments because they do not involve ambiguity and, thus, speed up decision-making processes. This refers to the Aristotelian paradigm (cf. Lakoff, 1987; Taylor, 2003). Yet, the analogic world demands the acknowledgment of nuances. The Aristotelian paradigm does not allow exceptions, lacks flexibility, does not account for contexts, and, ultimately, risks marginalizing the "different." What frustrates scholars (like me) invested in identifying ways to prevent classification from leading to discriminations is that classification is very appropriate in several contexts of our life. I would say that classification is needed in most contexts of our life. Bowker and Star (2000) discuss broadly the International Classification of Diseases (IDC), a methodology built by the World Health Organization (WHO) to identify diseases and support diagnoses and cures. The WHO defines the goals of the methodology as follows:

> ICD serves a broad range of uses globally and provides critical knowledge on the extent, causes and consequences of human disease and death worldwide via data that is reported and coded with the ICD. Clinical terms coded with ICD are the main basis for health recording and statistics on disease in primary, secondary and tertiary care, as well as on cause of death certificates. These data and statistics support payment systems, service planning, administration of quality and safety, and health services research. Diagnostic guidance linked to categories of ICD also standardizes data collection and enables large scale research.[1]

In summary, if you hope for a cure for cancer, want to see an orthopedic surgeon rather than a cardiologist if you break an ankle, or want to receive a vaccine before teenagers do (because you are 80 years old)—well, in all these cases, you should be very supportive of classification. In

[1] https://www.who.int/standards/classifications/classification-of-diseases.

fact, related to the latter example, classification played a prominent role in vaccine administration during the coronavirus disease of 2019 (COVID-19) pandemic. And failing to classify people with respect to their health needs can become an issue. For instance, in the United Kingdom (UK), scandals arose when evidence surfaced of people with "invisible" diseases being denied a fast lane to be vaccinated—something that had to do with an incomplete classification of their clinical picture. Cases, reported by *The Guardian* in 2021,[2] include Charles, a 26-year-old man from Coventry, whose blindness did not directly signify him as requiring a COVID-19 vaccine as a priority since it did not directly relate to red flags associated with the disease (such as respiratory issues, age, etc.). However, as Charles stated: "Despite not being considered clinically extremely vulnerable (CEV), I am at high risk due to needing assistance shopping, and my lack of depth-perception, which makes it hard to socially distance." Another example includes Frances-Mary Pratt, a 46-year-old woman from West Oxfordshire, whose diagnosed mild autism did not qualify her to receive the COVID-19 vaccine before otherwise healthy people over 50 years old. Her mental issues prevent her from having a normal life. As she reported: "I do not go into shops any more as there is a chance I shall be unfairly challenged for not wearing a mask. I am exempt from doing so, and wear a sunflower lanyard to prove it. However, there is always someone who won't accept it and challenges me unfairly."

These two examples illustrate the consequences of being superficial when classifying diseases or being careless of factors associated with a specific disease. Overall, one would recommend more in-depth classification of patients in order to make sure that the National Health Service (NHS), the UK's public funded healthcare system, can rely on complete and accurate datasets. But…classification is also "bad" as it leads to put human beings into categories, and this classification can put people at risk of being discriminated against, such as when, during apartheid periods, the South African Government decided to classify citizens by their ancestry, skin color, caste/class, and so on. Classification poses clear tradeoffs, and much depends on who uses classifications and for what. For instance, we should classify people's disease but should refrain from classifying one's sexual orientation. More specifically, healthcare systems and not private companies should classify people's disease (companies should

[2] https://www.theguardian.com/society/2021/feb/12/we-feel-forgotten-high-risk-groups-missed-off-uks-vaccine-priority-list.

probably not even be aware of personal health issues of their employees unless needed to create accommodations). But this argument has two caveats.

First, the fact that doctors, hospitals, and so on "mine" people's health data is good for some situations (e.g., vaccine prioritization above) and can actually advance medical research. But privacy issues surface. For instance, in the US, where healthcare is mainly private, knowledge of an insured person's preconditions might affect the person's monthly payment (premium) to healthcare maintenance organizations (HMOs) (also known as healthcare insurance companies) and the person's eligibility to receive healthcare insurance at all. To this end, an insurance company might deny a "diverse" individual (e.g., due to autism) healthcare due to the granular way in which it classifies diseases. It's like if Frances-Mary were asked if she'd rather be vaccinated but at the risk of losing her health insurance. Fortunately for her, Frances-Mary lives in the UK, which sees healthcare as a right rather than a privilege, so she won't have to deal with such a choice. Here, it is important to consider how information on classifications propagates throughout an invisible chain of command that involves institutional players such as insurance companies (as in the US).

Second, even classifying sexual orientation can be meaningful to promote diversity, equity, and inclusion (DEI). For instance, in 2021, the US census conducted a survey targeting the LGBTQ+ community[3] and discovered (comparing findings with other populations) that a much higher rate of transgender people suffered depression during the COVID-19 pandemic. This finding can be used in two ways. On the one hand, the study can serve as the baseline for advocating for more spending on mental health specifically addressed to certain communities (LGBTQ+). On the other hand, the study may reveal that the LGBTQ+ community will cost more tax dollars to citizens because its members are more prone to mental health issues and, thereby, more likely to access government-funded mental health support infrastructures. The former way to view these findings may lead to positive consequences. The latter can be used by the "system" to prevent the LGBTQ+ community from accessing government funds.

[3] https://www.census.gov/library/stories/2021/11/census-bureau-survey-explores-sexual-orientation-and-gender-identity.html.

All these examples illustrate the ambivalent role of classification. We cannot avoid classifications. It is the baseline of social order, yet who uses classifications and how determines positive or negative effects on businesses and society. Moving away from healthcare, in most countries, people pay taxes according to a progressive system (generally a tax bracket system) where a range of income is taxed at a given rate. They are "classified" on the basis of their income and "put into brackets" to tax them in a fair manner. In fact, when governments fail to classify individuals on the basis of their income, we generally refer to this as an *unfair taxation system*. In organizational settings, the same principles apply. Businesses need so-called organizational charts, which refer to the chain of command and reporting systems and which represent the baseline for compensation—and power (Galbraith, 1977; Mintzberg, 1983). Without applying social order to organizational life, there will be no organizations. But even these two taxation and organizational charts examples have pitfalls. For instance, in the US, tax brackets are arguably unfair as they are not granularly progressive (there are too few brackets). In addition, other types of taxes such as sales taxes hit all individuals in the same manner, which hugely penalizes people with low income. In a similar way, classifying employees might simply allow powerful people to deny promotions to women and Black people. The tax classification is relatively objective and might be wrong because the wrong metrics are used. The employee classification is subjective, not written down into organizational documents, and more in the minds of executives and top management. Yet, this type of classification based on gender and race that penalizes minorities is very much present, for instance, in the culture of high-tech companies in Silicon Valley.[4]

Classification in organizations and institutions gives powerful people technologies that can discriminate against minorities and vulnerable populations. But classification technologies can also be turned into ways to pursue social justice, such as by asking millionaires to pay more taxes than poor people and allowing women to take on leadership roles in business settings (e.g., the "pink quotes" in Italy, an initiative aimed at filling state-owned organizations with women in their boardrooms[5]).

[4] https://www.theatlantic.com/magazine/archive/2017/04/why-is-silicon-valley-so-awful-to-women/517788/.

[5] https://www.wsj.com/articles/SB100014240527023033956045774318321611 33916.

Algorithms and Classification

Why algorithms exacerbate classification is the *fil rouge* of the whole book. While, in the chapters that follow, I provide examples that are specifically tied to some technologies or business processes, here I set the stage by offering several meaningful examples of classifications and associated algorithms at work.

The first general example concerns gender and relates to software design. Being "binary" is a characteristic of digital systems (0/1). But that doesn't mean that the world should be digitalized in black and white. Or with just a few colors. While digital photos can be perfect replicas of the "film photography," a process that uses chemical processes to capture an image on a paper or film (they are provided with an almost infinite array of colors), digital details of human beings are often limited to mainstream ones. Until recent times, it was not uncommon to find online forms (job applications, surveys of various kinds, tax documents, etc.) with male and female as the only two options to declare one's *sex*. With the acknowledgment that sex and gender should be decoupled (Eckert, 1989) and the recognition that biological sex may not determine how someone wants to live their life (Bridges, 2011), online forms now commonly feature numerous options (*male*, *female*, *non-binary*, *transgender*, *prefer not to say*). Is that enough? In a recent *MIT Technology Review* article, Everett Franchuk, a queer writer on LGBTQ topics, pointed to the lack of options being offered to Gmail users when asked to provide their gender.[6] She points to the "other" option offered to users who do not fall into the binary system of biological sex and defined "other" as a dehumanizing term. I could not agree more. In addition, Everett pointed to the very simple technical workarounds required to address this uncomfortable issue and showed lines of code corroborating their point.

> It should not be difficult for companies to improve gender inclusivity on existing forms. Providing additional options often requires just changing or adding a few lines of code. Here's what it would look like to add a third-gender category in PHP, which is used to program many web forms.

[6] https://www.technologyreview.com/2022/08/22/1057853/software-coding-more-gender-options/.

This example is helpful to make two important points. First, even if it is true that machines strive for classification—or, better, they don't work without it—it is possible to make classification more ***diversity friendly*** with little effort (see Fig. 1.1). Second, one worrisome issue surfaces: even large companies with plenty of time and resources to write code that accommodates diversity don't do so. This, in my opinion, stems from a well-known issue affecting software developers in high-tech companies, that is, the lack of diversity in development teams. Research on team diversity is surprisingly consistent in asserting that the more diverse (gender, race, background) a team, the more inclusive its outcomes (cf., for instance, Barocas & Selbst, 2016; Cobb-Payton & Berki, 2019; Howard & Borenstein, 2018). Here, it is important to work in the direction of enabling teams of developers to acknowledge that people's attributes such as gender (I discuss race later in this chapter) are socially constructed with the aim of ordering the world or "sorting things out" (Bowker & Star, 2000)—a form of control. While we've luckily already moved away from a biological conception of gender (i.e., sex), gender issues emerge when people (or machines) attempt to label individuals according to their own mental frames. This is why team diversity allows for different frames and can, potentially, accommodate more individuals.

But, as Dr. Donna Maria Blancero, a professor of management and DEI (and former Provost) at Bentley University, told me in an interview for this book:

> [What is important] is the notion of cultural intelligence. Because people often say, if you have a diverse team, they're going to be more creative. That's not true. If you have a diverse team, short term, it's going to be chaos because they have to learn how to work together. But if you have a diverse team that's trained in understanding yourself, your own cultural values, and the values of others, then the productivity, innovation, et cetera,

```
<select name="Gender">
  <option value="">Select...</option>
  <option value="M">Male</option>
  <option value="F">Female</option>
  <option value="N">Non Binary</option>
</select>
```

Fig. 1.1 Code generating gender options (*Source* MIT Technology Review)

> goes up. … it's about educating people about differences. You're asking me at the right time because I'm covering this in class about how we all have our own cultural values. Those cultural values can come from our country, from our organization, from our ethnicity, even from our home culture. But we do things a certain way. I may be very short-term focused and you may be very long-term focused. It doesn't mean I'm right and you're wrong, and it doesn't mean you're right and I'm wrong. It means we're different. Until you and I can actually talk about that difference and monopolize on it and say, You know what? We can work together. You have the specialty, Donna, to think about short-term issues.

Her point is a good one. We need to pursue team diversity because everyone should have a voice in how systems of classifications such as AI systems are designed. Yet, we should not take for granted that a diverse team works as expected right away. Widder and Nafus (2023) recently made a complementary point concerning team diversity: "Divisions of labor, an important purpose of modularity, create the cracks through which responsible AI actions fall" they argue (p. 5). The argument being made here is that modern AI systems are very modular with several different sets of expertise needed to build the (complex) system that is then commercialized. The AI production supply chain, to this end, risks working in silos where different teams working on modules do not communicate with each other with the consequence that no one (team) takes responsibility to ensure responsible design of AI. The lack of a holistic view of a complex AI system prevents it from being built with embedded ethical principles. The comparison of AI software development with a supply chain is a good one. Supply chain systems are hardly integrated because they encompass different (often competing) companies with conflicting interests. There are interdependencies, but the fact that supply chains are formed by several different organizations make these interdependencies opportunistic and loose. In a similar way, the supply chain of the development of AI systems leads to individualistic behaviors where each team focuses on their specific part of the development process while missing the big picture, which concerns how the system as a whole will function when deployed in organizations and society.

But, at least with respect to team diversity, generative AI has the potential to overcome the lack of team diversity, something that, as I pointed out in earlier work (Marabelli et al., 2021a), is perpetuated by Silicon

Valley high-tech companies (consider the Gebru case,[7] discussed later in Chapter 5). One such example is how ChatGPT, one of the prominent generative AI software, works. While programmed by teams that might not be diverse, it extracts information from the Internet. The Internet is NOT the quintessence of diversity (cf. Bender et al., 2021). It does provide, however, some sort of democracy of thoughts. Below is the output of me asking ChatGPT a tricky question about gender in June 2023 (Fig. 1.2).

ChatGPT had just confirmed the relevance of the "wisdom of the crowd," which represents the collective opinion of a diverse independent group of individuals rather than that of a single person (Yi et al., 2012). Interestingly, the wisdom of the crowd overcomes teams' lack of diversity when they rely on broader networks such as the Internet.

But, that's not enough. ChatGPT (and generative AI more generally) is not yet ubiquitous—and in other contexts it can well discriminate.[8] When this happens, the consequences are extremely problematic. I next

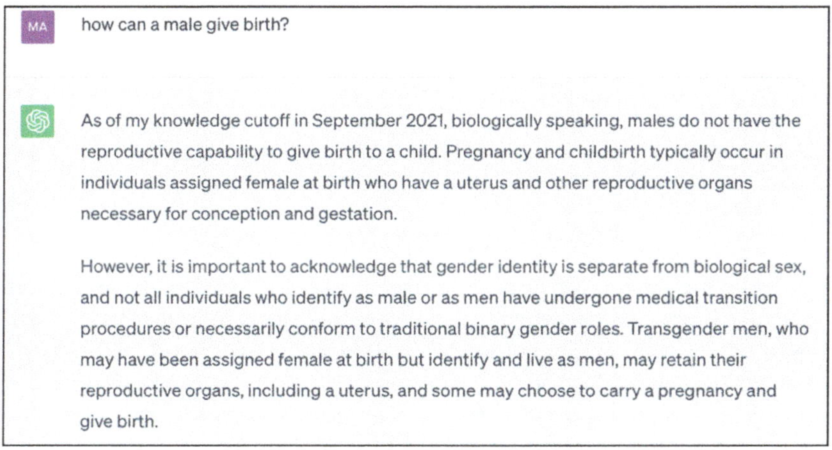

Fig. 1.2 Marco challenging ChatGPT

[7] https://www.nytimes.com/2020/12/03/technology/google-researcher-timnit-gebru.html.

[8] https://www.shrm.org/resourcesandtools/hr-topics/behavioral-competencies/global-and-cultural-effectiveness/pages/how-chatgpt-could-discriminate-against-applicants.aspx.

provide five key examples that I found disturbing and that involve classification issues. The first concerns the UK education system, the second concerns the development of fast lanes for having COVID-19 jabs administered and consequences for minorities (e.g., Black Americans), the third describes how hiring software classifies job candidates and discriminates against people with disabilities, the fourth discusses an opaque (and racist) algorithm that decides who, among indicted people waiting for trial, can be bailed out and who cannot, and the fifth concerns the US credit system and how citizens are granted (or not) a loan.

As already pointed out, classification is a required practice that permeates our everyday lives. A social system without classification is like a country without government. Immanuel Kant defines anarchy as a form of government relying on "laws and freedom without force." In a social system and with classification in mind, the equivalent of the Kantian's anarchy would be people paying taxes on a voluntary basis and in the amount they think is fair with respect to their income. In trying to avoid lashing back too heavily at Silicon Valley companies (successfully) attempting to offshore their taxes in more or less remote countries, I provide just one, representative example. In 2019, Facebook logged US$15B in profit in Ireland, which equates to US$10M per Facebook Irish employee (Desmond, 2023). No need for further comment on the extent to which entities (whether private citizens or companies) would feel responsible to pay "fair" taxes without laws and regulations.

The UK Education System

The first example concerns education systems, which have historically relied on classifications, at any level—from kindergarten to PhD programs. In early grades, students are labeled as good and bad students (nowadays, more or less performant students). Scholastic assessment tests (SAT) are needed in the US for college admission purposes. Graduate management admissions tests (GMAT) and graduate record examinations (GRE) are required for admittance into doctoral curricula in several countries. With the widespread introduction of algorithms, these education-related classifications were taken to the next (unhealthy) level.

Obviously, classifications that pervade our everyday life don't just happen in America. For instance, in 2020, in the UK, an algorithm was deployed aimed at grading students automatically (because of the COVID-19 pandemic that prevented students from attending school regularly) on the basis of a teacher assessment of individual students (rather than through national exams) and other information concerning the particular school in question. Since final grades were calculated including schools' historical performance, students from poorer areas were twice as likely to have their results downgraded than those from richer areas.[9] While, in general, students do worse in less affluent schools, including historical school data in the algorithm in this way meant it was impossible to account for individuals or small groups of students who might have been better than their historic peers. This example illustrates how one can conveniently use algorithms to represent and predict real-world phenomena—here involving children's performance in school. But because algorithmic classification relies on historical data (see Chapters 2 and 3 on data extraction and training), the completeness of the data (in this case) determines their likeliness of being truthful. In particular, this example showcases how struggling populations become more miserable when affected by algorithms that mercilessly ignore social struggles.

Classification is very often convenient for who classifies yet problematic for certain categories of individuals, which leads to unequal treatment of human subjects—in this example, poor students who were more likely to be penalized, never mind if they did well in the evaluation year. This is a diffused pattern in the AI realm where the designers of algorithms are efficiency-driven regardless of whether some people are left out. Pratyusha Kalluri (2020) explains very clearly how these power dynamics unfold. This is also an example of how the COVID-19 pandemic exacerbated inequalities, with the risk that these inequalities (often stemming from new practices, such as the automatic assessment of student performance based on historical data) perpetuate even as the global pandemic fades as I noted in a recent paper (Marabelli et al., 2021b).

Healthcare Injustices

Moving from education to healthcare, a second example of the pervasive and problematic use of classification concerns COVID-19 vaccination policies in various countries. Building on and expanding my prior example of UK vaccination policies and prioritization, in April 2021, the Joint Committee on Vaccination and Immunization (JCVI) identified nine

[9] https://www.thescottishsun.co.uk/news/5888751/sqa-results-downgraded-marks-poor-rich-deprived/.

priority groups for vaccination, including people over 50 years old, frontline health workers, social care workers, and people aged 16–64 with serious health conditions. However, because the national database managed by the NHS was incomplete, several people with serious conditions didn't receive the invitation to get the jab once the first vaccines were made available in 2022. In addition, people whose weight or ethnicity were not recorded in the database were automatically considered obese and Black (at that time, considered the highest risk ethnicity) and received the invitation to get their jab.[10] The classification system in the NHS database was problematic, which materialized in life-threatening scenarios. Here again, however, classification was needed because, once the first vaccines were approved by national health agencies, there was not enough supply for everyone. Prioritization required classification. Classification required looking into people's health record. Those not in the database were penalized as they could not receive the jab despite being eligible (some elderly). Others were unfairly prioritized because they lacked a complete medical record. Classifications systems processed by algorithms created inequalities, discriminated against some populations (e.g., the elderly), and processed data in ways lacking clinical evidence—Black people are not particularly at risk to get sicker than others when they contract COVID-19. But more Black people die to COVID-19 due to their statistical lack of access to health service. This issue is particularly evident in the US with studies confirming the relationship between Blacks' lack of access to care (testing, prevention, cures) and negative outcomes associated with them contracting COVID-19.[11] To this end, a bit of historical context of US discriminations in healthcare is needed.

Access to care has historically been problematic in the US for everyone but specifically for certain populations at the margin (e.g., the poor, Black communities, and undocumented individuals). But research has recently focused specifically on Black communities—probably also in light of the infamous Tuskegee Syphilis Study,[12] which lasted over 40 years and involved "experiments" on nearly 400 Black individuals affected with syphilis and purposely not treated with well-known, effective drugs to

[10] https://www.theguardian.com/society/2021/mar/09/scientists-question-nhs-algorithm-as-young-people-called-in-for-jab.

[11] https://www.ncbi.nlm.nih.gov/pmc/articles/PMC7762908/.

[12] https://www.cdc.gov/tuskegee/timeline.htm.

observe the consequences of the disease if not treated. This and other unethical research misconduct led to the Belmont Report,[13] which spells out ethical practices when conducting research involving human subjects. Even if abuses such as those happened during the Tuskegee Syphilis Study do no longer happen, things have not changed substantially in recent times, with respect to treating Black American differently, with respect to access to care. A 2023 *Nature* article reports that Black and Hispanic people are up to twice as likely as White people to develop Alzheimer's disease, yet they made up no more than 20% of participants in trials for new monoclonal antibody drugs against the disease (Reardon, 2023).

Unlikely the UK example of working algorithms and flawed datasets, here discriminations stemming from classification focused on (and targeted) specific populations of minorities (in the US, Black people make up roughly 14% of the country's population, according to 2022 Census data[14]). Black Americans are often wrongly "classified" with negative attributes (i.e., that they are poorly educated, are more prone than others to commit crimes, and don't go to doctors), but these attributes do little justice to the underlying social systems. Let me dig deeper into these issues.

Black Americans do go less to college according to the US National Center for Education Statistics (NCES), 21% of Black adults earned a college degree versus 35% of White adults[15] (last time updated in 2019). Yet, college access has been historically barred for Black Americas by a number of issues (redlining policies,[16] Jim Crow laws,[17] etc.) that all fall under the umbrella of structural racism—critical race theory (CRT) being the main approach taken to illustrate how current laws and regulations still allow racism, at the systemic level (meaning, de jure, i.e., by not breaking the law) (cf. Bridges, 2019 for a comprehensive discussion of CRT). Twentieth-century redlining policies generated segregations, such as with how banks were encouraged by the Federal Housing Administration (FHA) to deny mortgages to Black people in particular "fancy"

[13] https://www.hhs.gov/ohrp/regulations-and-policy/belmont-report/index.html.

[14] https://www.census.gov/quickfacts/fact/table/US/IPE120221.

[15] https://nces.ed.gov/programs/raceindicators/indicator_rfa.asp.

[16] https://www.npr.org/2017/05/03/526655831/a-forgotten-history-of-how-the-U-S-government-segregated-america.

[17] https://www.nps.gov/malu/learn/education/jim_crow_laws.htm.

neighborhoods (residential suburbs, which White people generally inhabited), literally *redlining* city and town maps (Rothstein, 2017). This type of segregation has legally (but wrongly) taken place despite the 1866 Civil Rights Act[18] with which Congress enforced the abolition of slavery and prohibited actions that it deemed to perpetuate its characteristics. Two landmark decisions of the Supreme Court reversed this trend of not just structural but also institutionalized, yet illegal, racism—meaning that, even if the 1866 Civil Rights Act banned racist practices (practices that resemble those characterizing slavery), American institutions acted otherwise, promoting segregation, in the name of the "separate but equal" principle. The first landmark decision was the 1954 Brown v. Board of Education of Topeka[19] ruling whereby the Supreme Court ruled unanimously that racial segregation of children in public schools was unconstitutional (a cornerstone of the civil right movement at that time) and helped establish the precedent that separates but equal education and other services were not, in fact, equal at all. The second was the Jones v. Mayer[20] decision of the Supreme Court ruling in favor of Barbara Jo Jones, who sued the Alfred H. Mayer Company, which refused to sell her a home just because she was Black. Three years after, the US Congress adopted the 1968 Fair Housing Act, which "Prohibits discrimination by direct providers of housing, such as landlords and real estate companies as well as other entities, such as municipalities, banks or other lending institutions and homeowners insurance companies whose discriminatory practices make housing unavailable to persons because of race or color; religion; sex; national origin; familiar status; disability.[21]" This history of injustice prevented Black Americans from accessing the education system and, yes, made them prone to commit more crimes. In fact, according to the most recent FBI reports on crimes,[22] in 2019, the total arrest of Black Americans was 1,815,144. Normalized with respect to the US

[18] https://history.house.gov/Historical-Highlights/1851-1900/The-Civil-Rights-Bill-of-1866/.

[19] https://www.uscourts.gov/educational-resources/educational-activities/history-brown-V-board-education-re-enactment.

[20] https://supreme.justia.com/cases/federal/us/392/409/.

[21] https://www.justice.gov/crt/fair-housing-act-1.

[22] https://ucr.fbi.gov/crime-in-the-u.s/2019/crime-in-the-u.s.-2019/tables/table-43.

population, 26.6% of arrests were of Black Americans (Black Americans were 12.2% of the US population in 2019).

While the difference is not substantial, it is true that, statistically, *Black Americans were arrested more than other populations*. But, first, being arrested does not equate with being convicted. In fact, we need to account for predictive policing aimed at patrolling specific neighborhoods—even with algorithms (see Marabelli et al., 2021a) and stop-and-frisk tactics used in discriminatory ways (Gelman et al., 2007). Second, redlining and segregations generated "projects," or ghetto neighborhoods, where poorly educated, largely unemployed people were more prone to commit crimes. Is that a justification? Not at all. But clearly we cannot deny that systemic practices existed (and exist) that lead to situations that substantially disadvantaged Black Americans. Failing to understand these historical struggles of minorities and people at the margin would simply perpetuate structural racism, prejudice, stereotypes, and, ultimately, widen rather than fill the gap between privileged and unprivileged in our society.

Black Americans do go less to doctors too. Probably, just skimming the Belmont report would raise enough concerns—any Black American is entitled to be skeptical when thinking of the extent to which the US healthcare system may take care of them. But let's not forget that the US system is predominantly a private system. Public healthcare (so-called safety nets[23]) offer "some" service to uninsured and with inconsistent quality—Khiara Bridges, Law Professor at UC Berkeley School, provides a good reading about this in her work *Reproducing race: An ethnography of pregnancy as a site of racialization* (Bridges, 2011). She explains how public hospitals (and doctors and staff) classify individuals based on skin color, accent, legal status, and culture. This type of socially constructed classification (what is not a socially constructive classification on human beings?) leads uninsured individuals to access (public) care with skepticism and envoy—compared with privileged individuals who can access private care. Private care is not perfect too. In her popular book *Automating Inequality: How High-Tech Tools Profile, Police, and Punish the Poor* (2018), Virginia Eubanks, Associate Professor of Political Science at the Rockefeller College of Public Affairs and Policy at

[23] https://aspe.hhs.gov/topics/human-services/social-safety-net.

the University of Albany, explains how algorithms, relying on human-built categories, heavily discriminate against people of color, immigrants, and more generally people at the margin. These algorithms are used in public as well and private healthcare systems in the US. Issues with algorithms may happen because a database has missing data (e.g., the UK database's flaws), programmers made errors in designing algorithms, or simply because people in power design systems (algorithms, and the underlying classifications) that are aimed at preventing some people from accessing basic health services. Remember: healthcare systems become costly as people become sick. Therefore, a "well" (unethically) designed system would prevent people more prone to become severely ill to access the healthcare system. Once they are in, it is hard to deny them the care they need. Therefore, barring access to care is something that might do the trick. It is also something that has cost uncounted lives in the US over the years.

To summarize, it is true that Black Americans go less often to doctors than others, such as White Americans. While this is evident, still, clinical research is often based on statistics that do not account for access to care when attempting to classify the severity of a disease by race. If Black Americans are mostly barred from accessing care (whether because they are unemployed and, therefore, not supported by private care or because they do not trust the system or not able to afford care), can anyone say that they die for COVID-19 at a higher rate than Whites because of the color of their skin? Probably, running *dry* numbers and aiming at correlations, one may come to the conclusion that Black Americans were more prone to face adverse outcomes when contracting COVID-19. But the underlying historical facts confirm that correlation is different than causation. The correlation vs. causation tradeoff plays a big role in AI processes of classification. The example of Black Americans and COVID-19 is a good one. Healthcare systems classify people by race. They notice that Black Americans die at a higher percentage. Therefore, Black Americans have "something" in their DNA that makes them more prone to adverse consequences from COVID-19. That's the correlation turned into causation. WRONG. Black Americans die more to COVID-19 due to social issues associated with structural racism. Correlations are not enough to explain phenomena. Unfortunately, AI revolves around classifications but is not informed by enough variables to understand when correlations cannot explain cause-effect patterns (Luca et al., 2016).

Classifications are needed to preserve social order (and advance in medicine, for example). But classifications are also dangerous because they tend to label individuals and phenomena in a correlation-based fashion. For instance, Black Americans are often labeled as unhealthy, problematic, unlikely to go to college, prone to commit crimes, and so on simply because of their lack of opportunities to access resources such as healthcare, education, and housing (think of all redlining policies still in place, illegally,[24] which perpetuate segregation of minorities).

Jyoti Madhusoodanan raised an interesting yet disturbing insight on classifications concerning healthcare and Black communities to the surface recently in a *Wired* (US edition) article in October 2021.[25] A quote from the outset of the article effectively summarizes the issue at hand:

> Recently, two leading medical associations recommended ending a decades-old practice among doctors: using race as one of the variables to estimate how well a person's kidneys filter waste out of their bodies. Before, clinicians would look at the levels of a certain chemical in blood, then multiply it by a factor of approximately 1.15 if their patient was Black. Using race to estimate kidney function contributes to delays in dialysis, kidney transplants, and other life-saving care for people of color, especially Black patients.

A similar example offered by Madhusoodanan refers to a risk calculator aimed at predicting the odds of dying a month after a positive COVID-19 test. This also related to access to care. The researchers (cf. King et al., 2020) reviewed COVID-19 testing results within veterans hospitals between February 8 and August 18, 2020, and applied logistic regression models considering demographics (age, sex, race/ethnicity) and preexisting medical conditions and the Charlson Comorbidity Index (CCI) derived from ICD-10[26] diagnosis codes. They found that race did not affect COVID-19 survival (in contrast with then mainstream research suggesting that race was a factor in negative outcomes from being infected by COVID-19). In reality, patients in the veteran hospital system simply experienced fewer barriers to accessing care. As Madhusoodanan suggests:

[24] https://www.npr.org/2017/05/03/526655831/a-forgotten-history-of-how-the-U-S-government-segregated-america.

[25] https://www.wired.com/story/health-race-medicine-risk-disparities/.

[26] https://www.who.int/standards/classifications/classification-of-diseases.

"With fewer barriers, the seemingly race-based difference in risk of death was minimized." Evidence of the fact that the initial findings by Joseph King and colleagues suggesting that Black people were more prone to die from COVID-19 did not account for access to care were further corroborated by a 2022 study conducted at the Harvard School of Public Health.[27]

Khiara Bridges (2011) described broadly the pitfalls of relying on biological characteristics to make sense of certain medical conditions in her ethnography of how pubic hospitals operate under the US Medicaid program (a government program, established in 1965, that provides health insurance to citizens and permanent residents—with some exceptions made to non-permanent residents—with limited income and resources). Science questions most assumptions that race should influence cures to patients. Socioeconomic and cultural factors, frequently used to discriminate against populations, are often the most likely explanation for why some "races" are more or less affected by certain diseases. Probably, one of the most astonishing examples is that of HIV in the 1980s, when most people (even eminent clinicians) began to assume that the disease could be passed only (or most likely) via sexual encounters between men.

Nowadays, it is well-known (read: *it's Science*) that HIV is not a "Black" or "homosexual" disease (or "gay plague" as it was first named[28]). Yet, stigmata seem to stick with the past. Having recently come across a commercial of the HIV drug Apretude[29] (aired in the US in 2023), I was under the impression that there was a heavily prevalence of Black and gay people in the commercial. Therefore, I conducted a 10-minute analysis of the actors and actresses in the commercial and show the results in the Table 1.1.

It is true that, still, nowadays, Black Americans are more prone to contract HIV than other populations and, thereby, represent the potential "target" of HIV medications. However, as Table 1.2 shows, White people represent 28.5% of the HIV population (2015–2019). Given that within that timeframe the HIV population in the US was around 1.2 million people,[30] 342,000 White people had HIV (compared with 483,000 Black Americans). Does this difference justify a commercial targeting specifically

[27] https://www.hsph.harvard.edu/news/hsph-in-the-news/covid-death-rate-now-higher-in-whites-than-in-blacks/.

[28] https://www.nbcnews.com/feature/nbc-out/three-decades-later-men-who-survived-gay-plague-speak-out-n825621.

[29] https://apretude.com/.

[30] https://www.hiv.gov/hiv-basics/overview/data-and-trends/statistics/.

Table 1.1 Apretude Commercial on HIV Drug[31]

seconds	description	classification	seconds	seconds	white
1-3 sec	white male smiling	white			other
3-9 sec	two black males kissing	black, gay			
9-14 sec	black gay guys dancing	black, gay			
15 sec	black woman	black			
16 sec	white woman	white			
17 sec	black man	black			
18-23 sec	white and black men kissing	interracial, gay			
24 sec	club with black men (gay club)	black, gay			
25-28 sec	black man smiling	black			
29-37 sec	black eterosexual couple	black			
37-40 sec	three men smiling at each other (black and white males)	interracial, gay			
41-45 sec	black woman first with girlfriends then alone	black, gay			
46-55 sec	two gay men (one black, one white) looking at each other at a train station	interracial, gay			
56-59 sec	a white and a black man kissing	interracial, eterosexual			
59-00 sec	Apretude commercial	product	4	55	

Black Americans? Or is this a way to perpetuate a longstanding stigma? In defense of Apretude, one might argue that, because Whites have broader access to care (and related information on diseases) it is Black communities that need to be targeted (with television ads). In this way, Apretude would provide a service to minorities that might not be aware of cutting-edge medications to manage HIV. I am not sure that this theory would fly among Black Americans, but it is something worth mentioning.

Table 1.2 Impact of HIV on racial and ethnic minorities in the US[32]

Race/Ethnicity	% of People with HIV, 2019[11]	% of U.S. Population, 2019[14]
Black/African American	40.3%	13.4%
White	28.5%	60.1%
Hispanic/Latino	24.7%	18.5%
Asian	1.5%	5.9%
American Indian/Alaska Native	0.3%	1.3%
Native Hawaiian and Other Pacific Islander	0.09%	0.2%

Source CDC, Estimated HIV incidence and prevalence in the United States, 2015–2019, HIV Surveillance Supplemental Report 2021; 26(1) and US Census Bureau, Quick Facts–United States

[31] https://www.youtube.com/watch?v=KRozChEd1ww.

[32] Source: https://www.hiv.gov/hiv-basics/overview/data-and-trends/impact-on-racial-and-ethnic-minorities/.

Overall, and regardless of the Apretude commercial (it is just an example among many others, and actually their contribution to HIV is commendable), I am under the impression that the HIV stigma applied to Black Americans in the 1980s is still a living thing after 40 years. In addition, reports of more recent practices such as the examinations of one's kidneys' function are examples of classification applied to race (but also to "populations." Bridges takes the discussion on populations further and highlights that population "works by constructing what appears to be a self-evident group around a selected characteristic, feature, or attribute. The groups enacted by population conceal their social construction insofar as, once constructed, the populations appear to be "natural"; that is, they appear to be natural ways to refer to social phenomenon" (Bridges, 2011, p. 179). The notion of population is extremely relevant for classification processes whether enacted by human beings or machines, such as AI. Population can be hardly fed "manually" into an algorithm because, as Bridges indicates, it is socially constructed and, thereby, emerges through social practices. However, given that nowadays most AI systems are powered by machine learning (ML) engines, it is almost obvious that such models will embed populations simply by observing, mining, processing, and replicating human behaviors and beliefs. This currently happens when ML scrape the Web (social media and other platforms such as Reddit). Population embedded in AI systems can easily transcend race, or stigma, which can make classifications even more discriminatory yet hardly noticeable to most people (and, therefore, difficult to address with laws and regulations). For instance, Bridges (2011) shows how physicians classify patients not just on the basis of their race but also with respect to behavioral characteristics of some populations. For instance, a doctor interviewed by Bridges reported that Chinese populations "have a lower level of tolerance for difference and disability" (p. 138), while another doctor asserted the "Bangladeshi crowd" don't turn their cells off when they are in the hospital. The same doctor indicated that "Spanish, the Hispanic crowd are much more compliant to that kind of thing" (p. 128). Bridges labeled this as *cultural racism*, a way to discriminate against minorities in an even sneaker manner than "pure racism," the former being invisible, spotted only by mining behaviors, scrutinizing accents, and even looking into one's immigration status or place of birth. All this makes classification revolving around cultural racism harder to address and condemn. It is based on stereotypes and generalizations that are tied to the very characteristics of populations and historical events. When all this becomes data processed by algorithms, it is not hard to understand why biases embedded in AI are very hard to correct. An explanation, but not a justification.

Automated Hiring Software

Automated hiring software are powered by AI systems and have the goal to minimize the time (for recruiters) to identify a shortlist of candidates. In fact, for most jobs offered by large companies, there are thousands of applicants. To reduce the time-to-recruit and to achieve efficiencies, companies often rely to software able to "skim" resumes and discard unfits. For instance, indeed,[33] a worldwide hiring company, "pulls information from resumes for recruiters using something called natural-language processing, which involves extracting relevant words and phrases from text using computer programs."[34] These recruiting types often lead to candidates gaming the system (Kellogg et al., 2020), such as by injecting specific keywords in their resumes such as "execute," "reliable," "success," and so on—and there are websites that specifically recommend how to do so.[35] This is problematic because it advantages people with a specific knowledge of algorithms and how to game them. For instance, think of two applicants for a cashier job at a grocery store. The first one is a young millennial who spends most of their time online and who has learned about how algorithms work when scanning resumes. The second is a boomer, just laid off from their previous job, who wants to land a new job for a few years until retirement to make ends meet and meet the requirements for being eligible for pension (and Medicare in the US—the government-funded health insurance for people 65+ years old). The millennial will be able to craft a resume that is appealing to AI recruiters. The boomer will miserably fail just by crafting an honest application. Note that, in the US, it is illegal to discriminate job applicants by age. Yet here AI acts as a proxy for discrimination.

But what is more worrisome is the next step, the interview. AI plays a pivotal role in interviewing candidates. These systems are trained to

[33] https://www.indeed.com.

[34] https://www.forbes.com/sites/karstenstrauss/2018/02/02/the-role-of-artificial-intelligence-in-the-future-of-job-search.

[35] https://blog.sprintcv.com/the-importance-of-keywords-in-your-cv-how-to-optimize-for-applicant-tracking-systems/.

capture feelings and emotions using facial recognition software—even though such software is notoriously bad at capturing people's feelings (Crawford, 2021b). Also, people with disabilities are often flagged as not suitable for any job when an AI chatbot interviews a candidate. This happens simply because these individuals do not conform to what is believed to be "normality" by the system—clearly fed with data that does not account for, for example, an individual with autism (Sánchez-Monedero et al., 2020). In her recent book *The Quantified Worker* (Ajunwa, 2023), Ifeoma Ajunwa, AI.Humanity Professor of Law & Founding Director AI and the Law Program at Emory University, highlights how American workers are reduced to numbers by AI technologies in the workplace, automated hiring systems being the most prominent. These systems classify applicants according to a score or rank, often ignoring the gestalt of their human experience. Sometimes, Ajunwa argues, they even sort people by their race, age, and sex—a practice that's legally prohibited from being part of the employment decision-making process.[36]

In America, we are famously behind when talking about individual rights to fight algorithms. This has its roots in sad events such as 9/11, after which the Patriot Act[37] legitimized the US Government to oversee private conversations between citizens with expert data-mining systems (among many other government actions that are still controversial from a privacy standpoint). After several sunset provisions (measures that provide for laws to cease to be effective after a specified date), the Patriot Act ended officially in 2020.[38] We are behind if compared to Europe for instance (e.g., the AI Act,[39] which might propagate worldwide as it regulates several Internet services on data collection). But the good news is that, with respect to the role of algorithms and AI in hiring people, in New York City, new legislation is being rolled out to determine how much

[36] For a summary of the book's insights, you can refer to the author's article appeared in Wired Magazine in May, 2023, and accessible through this link: https://www.wired.com/story/unfair-automated-hiring-systems-are-everywhere-ftc/.

[37] https://www.fincen.gov/resources/statutes-regulations/usa-patriot-act.

[38] https://www.nytimes.com/2020/03/27/us/politics/house-fisa-bill.html.

[39] https://www.europarl.europa.eu/news/en/headlines/society/20230601STO93804/eu-ai-act-first-regulation-on-artificial-intelligence.

AI can influence job applications.[40] This is a small step for humans but a big leap for a country where the power of algorithms, in the past decade, has been in the hands of large companies in an exclusive way. This bill requires more transparency in how companies scrutinize applicants and make decisions yet embeds important caveats. A summary of the law is reported on nyc.gov:

> This bill would require that a bias audit be conducted on an automated employment decision tool prior to the use of said tool. The bill would also require that candidates or employees that reside in the city be notified about the use of such tools in the assessment or evaluation for hire or promotion, as well as, be notified about the job qualifications and characteristics that will be used by the automated employment decision tool. Violations of the provisions of the bill would be subject to a civil penalty.

One issue concerns how transparency is enforced. The bill requires companies making use of AI to hire personnel to undergo regular auditing processes (by a third-party company). These auditing companies are supposed to scrutinize the algorithms being used to select candidates and make an assessment on whether such algorithms are, in some way, discriminating in ways that are against the law. For instance, in the US, it is illegal to discriminate against someone based on race, gender, age, and so on. However, as it is pointed out by Ben Winters, the lead person of the AI and human rights project at the Electronic Privacy Information Center (EPIC[41]), the fact that an auditing company doesn't find problems in an algorithm doesn't mean that the algorithm doesn't have problems.[42] In other words, because this bill is the first of its kind and the technology at hand is relatively new, it will be easy for companies to find ways to bypass audits (and be certified as transparent and fair). Overall, while I fully understand the convenience of using automatic systems to skim resumes of thousands of applicants, I am skeptical when reading of companies making a heavy use of such systems. As per many other aspects of AI and algorithms, one faces tradeoffs between efficiency on the one hand and fairness and ethics on the other hand when using

[40] https://legistar.council.nyc.gov/LegislationDetail.aspx?ID=4344524&GUID=B051915D-A9AC-451E-81F8-6596032FA3F9.

[41] https://epic.org.

[42] https://qz.com/americas-first-law-regulating-ai-bias-in-hiring-takes-E-1850602243.

them. Certainly, AI for hiring has the potential to discriminate across populations, which would lead to the exclusion of certain categories of individuals and, thereby, equality issues.

Juridical System

In 2018, an algorithm (called COMPAS) was created that supports several US court decisions for bailing out (or not) suspects awaiting trial charged with minor crimes (Courtland, 2018). The algorithm was meant to speed up judicial decisions so that people eligible for bail could benefit from it and detention centers could become less overpopulated (this being a longstanding US issue). In most cases, the algorithm—based on the "best practice" of examining people's background, along with other metrics—was successful. However, it was demonstrated that, in some instances, the algorithm had substantially penalized Black Americans for various reasons, such as the fact that statistics on criminals at risk of fleeing before the trial include considering the area in which they live. But neighborhoods are often ethnically characterized. Put simply, Black applicants for bail might be penalized because they live in particular neighborhoods and, therefore, according to the algorithm, were more at risk of not showing up for trial or were more prone to commit a(nother) crime before being tried. Using automated software to make decisions on the lives of human beings is something that should not happen whatsoever. As Rhema Vaithianathan, Professor of Health Economics and Director of the Centre for Social Data Analytics at Auckland University of Technology, suggests, it is important to *always* have a human in the loop, which means that some sort of human supervision needs to be done when the fate of individuals are at stake. For instance, the EU is building AI laws that account for the extent to which an AI system directly impacts human beings and, thereby, implicitly assessing the risk that some AI outputs can lead to negative consequences (European Union, 2021). For instance, an AI managing the production (or even commercialization) of cars needs to be overseen but a miscalculation of the AI systems will only lead to slowing processes or selling less cars (or, in the worst case, misrepresenting the value of a car to perspective customers). This type of AI needs supervision, but the human does not need to be in the loop *all the time*—doing otherwise would simply blow the whole purpose of automating processes. The next level is supply chains, which notoriously rely on algorithms and AI because they generate simulations and digital twins/scenarios—as I recently discussed in a podcast with a friend and supply chain and operations analytics leader at Accenture.[43] Supply chains involve the movement of goods now happening worldwide. Especially when trading perishable products, it is paramount to generate timely supply chain movements. Here, predictive analytics and AI are key to forecast consumer needs. At this level, a human in the loop is needed because failure to supply grocery stores might affect individuals. What if a supply chain system crashes and people don't find food on supermarkets shelves? In the supply chain context, the human in the loop is required because it is critical for human beings to receive fresh products to purchase. But here, too, some supply chain processes can be automated with little to no human supervision. However, processes such as hiring need *constant* human in the loop supervision for two main reasons. First, compared to line production or supply chains, hiring processes led by AI can lead to (unintended) discriminations, which will jeopardize the present and future of people (most likely, people who are classified as "different"). At least, with a supply chain, failing to supply a grocery store affects people in the same way (in principle, as wealthy people have always more resources to "get by"). Second, applicants need feedback when rejected. And AI systems are known for not being able to provide feedback to humans (Tarafdar et al., 2022).

Bottom line, the example of the juridical system adopting algorithms showcases, once again, how AI and algorithms classify individual in a very shallow fashion and, thereby, lead to dangerous discriminations. For instance, Black Americans who are fathers and mothers may be denied to come home while waiting for trial just because an algorithm, based on skewed data, decided that they might be more at risk than others to flee (often, the others are not people who are less at risk of fleeing, they are just Whites). This doesn't mean that backgrounds should not be checked or that Blacks should be privileged. This means that if an algorithm is racially biased, it needs to be corrected (and while this happens, it needs to be dismissed).

[43] https://podcasters.spotify.com/pod/show/bentley-university-mccall/episodes/Lessons-learned-from-the-supply-chain-disruption-during-the-COVID-pandemic-e25keaq/a-aa04uuu.

Access to Credit in America

The third and last example of problematic classification concerns access to credit. In the US, anyone who aims to access credit (credit card, car loans, mortgages, possibility to rent an apartment, etc.) as well as access government jobs and, sometimes, jobs at private companies and nonprofits need to have a credit score in good standing. One of the most used credit score agencies is the Fair Isaac Corporation (FICO), which defines credit scores as follows[44]:

> A FICO Score is a three-digit number based on the information in your credit reports. It helps lenders determine how likely you are to repay a loan. This, in turn, affects how much you can borrow, how many months you have to repay, and how much it will cost (the interest rate). When you apply for credit, lenders need a fast and consistent way to decide whether or not to loan you money. In most cases, they'll look at your FICO Scores. You can think of a FICO Score as a summary of your credit report. It measures how long you've had credit, how much credit you have, how much of your available credit is being used and if you've paid on time. Not only does a FICO Score help lenders make smarter, quicker decisions about who they loan money to, it also helps people like you get fair and fast access to credit when you need it. Because FICO Scores are calculated based on your credit information, you have the ability to influence your score by paying bills on time, not carrying too much debt and making smart credit choices. Thirty years ago, the Fair Isaac Corporation (FICO) debuted FICO Scores to provide an industry-standard for scoring creditworthiness that was fair to both lenders and consumers. Before the first FICO Score, there were many different scores, all with different ways of being calculated (some even including gender and political affiliation).

Turns out that, in America, if you don't have a "good" credit score, you are barred from accessing basic right such as work, housing, and so on—yet, classifying people according to credit scores is full of pitfalls. First, credit scores advantage people with jobs and particularly wealthy people. Credit scores not just allow individuals to access credit and housing. They also allow better deals to people with a better credit score. The higher the credit score, the lower will be your interest rate if you borrow money to buy a house. This makes sense for lenders as credit score serves as

[44] https://www.myfico.com/credit-education/what-is-A-fico-score.

an index to figure out the risk of landing money. But this, as very well explained by Desmond (2023), disadvantages people who are more in need to borrow money. This type of discrimination is all in bright light. You have money; therefore, you can pay your bills on time and, thus, can access credit at more convenient conditions. If you are covered in debts and late with payments, your ways to access credit change substantially. Because credit card companies can access (more or less) public records, these (predatory) companies tend to send offers in the mail to people with struggles. They offer credit cards with a relatively low line (one or two thousand dollars) at very high interest rates (justified by the low credit score of the potential applicants), but, more importantly, they ignore or hinder the borrower's ability to repay the debit. According to debt.org, a private company serving the public with accurate and accessible information online about financial wellbeing, predatory lenders typically target various types of minorities, the poor, the elderly, and less educated individuals. To wit, it is actually incorrect to assume that people with financial hardships cannot access credit. They can but in a way that will make them even more miserable because they will become insolvent after paying high interests to predatory credit companies (that will nevertheless make money from these practices due to high interest rates). And when these companies send the credits to collectors, the alternatives for the victims are to either file for bankruptcy or access credit using a credit system that, in parallel to the banking system, offers credit but with incredibly high interests. This parallel system is represented by the payday loan industry. Payday loans are short-term loans for a small amount of money generally due on the next payday of the person who borrows the money. In recent times, several US states have capped interest rates, which have made payday lending illegal. Yet, the practice is still current in some states and, if not, these systems run illegally.

Algorithms have a prominent role in all this. While credit scores such as FICO are supposed to be transparent (and anyone can obtain, for free, a yearly credit report from the three major credit score companies in the country), low (but decent) credit scores (around 600 out of 850 points) are seen by lenders as suspicious. In the past, it was relatively common practice for some banks to combine "traditional" data such as employment, income, and credit score with more sneaky pieces of information, such as applicants' online behaviors on social media that the banks collected and analyzed with automatic systems. As per a *Wall Street Journal* report, the way this information is used is unclear (leading to algorithmic opacity) and adds to the unfair mechanisms to access credit. In fact, these practices that are not transparent to customers do not give them a way to "fix the problem"—something that the EU addressed years ago with regulations mandating that lenders provide specific reasons when rejecting loan applicants and, thus, greater transparency into how applications are processed.[45]

The way people can access credit relies on classification practices that are often discriminatory and, in the US, exacerbate social differences. AI systems are being used to mine unconventional data such as online activities and even one's Facebook network of friends.[46] While some claim that this would give lenders more accurate data to make decisions on whether to approve requests from borrowers, these practices are highly nebulous and prevent people from figuring out why they were denied a loan, which makes it nearly impossible for them to improve their chances to access credit. The overall system raises equity issues. The more one is knowledgeable of how these mechanisms work, the more they will be able to improve their chances to obtain a loan, often simply by gaming the system (Kellogg et al., 2020). In America, people are classified by their financial behaviors. This classification system rewards people who are already advantaged because they have a good and well-paid job and because they have the knowledge and education to understand how to keep their credit score in good standing. An additional, hidden system maps data points that are unknown to the most (i.e., online behaviors). This practice further disadvantages categories of individuals who are already in financial trouble. To summarize, classifications made by lenders on borrowers simply widen differences between categories of individuals, helping the rich and letting down the poor. The introduction of sophisticated algorithms and AI to support the aggregation of data from disparate sources to make up one's ability to repay a loan adds to these ambiguities. One interesting thing is that these automatic systems are not really fully automated as loan managers can generally override the algorithms. For instance, if John Smith files an application for a mortgage, his credit score is good, he has a job, and he earns at least three times greater than the supposed monthly mortgage installment each month (a general rule of

[45] https://eur-lex.europa.eu/legal-content/EN/TXT/PDF/?uri=CELEX:32014L0017&rid=6.

[46] https://www.theatlantic.com/technology/archive/2015/09/facebooks-new-patent-and-digital-redlining/407287/.

thumb used by most lenders to approve long-term loans) but the system rejects his application, the loan manager has the ability to manually review it and decide to approve the loan anyway. It might have been the case that red flags spotted by AI systems on John Smith's social media behaviors are practically irrelevant. But, if loan managers decide to override the algorithms, the responsibility in case the borrowers don't repay their loans is on them. With this mechanism, most loan managers prefer to follow algorithms blindly to avoid consequences affecting their own job. In this way, there is indeed a system that includes a human in the loop. Yet, the reward/punish internal systems of most banks prevent the human in the loop to act as they should without the fear of consequences, which means they are pointless and ineffective.

Conclusions

In this chapter, I bring to the surface several ways in which people and algorithms classify people. Classifications are among the most common practices that occur in social life, but it is important to remember that classifications are socially constructed. This is very evident when we refer to an organizational chart or when governments create tax brackets. Classifications become more problematic when, for instance, we refer to healthcare classifications. The IDC system represents a good *modus operandi* to classify diseases—and we need such a classification. But these systems in healthcare risk leading to classifications concerning race and gender and might become discriminatory, which would further increase inequalities in our society. They might also exclude people from accessing basic health services. Implicit healthcare classifications, often originated by practices that do not mean to be racist, nevertheless lead to racist outcomes. The example of doctors assessing a patient's risk of kidney failure on the basis of race is one such example. But, during the COVID-19 pandemic, we witnessed many other examples that are similar in substance. Good intentions, bad outcomes. For instance, during the pandemic, cases emerged of health technologies that didn't work well for minorities, such as Black Americans. This stems from the fact that these populations have long had less access to care (Marabelli et al., 2021b). Even oximeters (over-the-counter devices that attach to the finger and provide blood oxygen reading and that became extremely popular during the pandemic) are designed for light skin tones and do not perform well for people of color (Moran-Thomas, 2020).

The introduction of algorithms and AI in our everyday practices exacerbates classifications. AI systems need to be fed with data to classify because this is how they work—in a binary way. Adding nuance to binary systems is possible if programmers are driven by goals concerning principles of inclusivity—but this is not something happening systematically. Large high-tech companies are not sensitive to these values as they are revenue-driven. And revenues underpin efficiency. And efficiency often involves sloppy programming, outsourced coding, and poor concern for potential negative societal outcomes. The end game is systems that use classifications for profit rather than for serving citizens fairly. Things are indeed changing. After the #metoo movement and the assassination of George Floyd, we have witnessed a worldwide growth of initiatives that focus on inequalities and on pursuing DEI principles, which stresses the growing demand for more social justice (Marabelli et al., 2023; Oswick & Noon, 2014). These principles, however, are often difficult to embed in automatic systems for myriad reasons that I explore in subsequent chapters. For one, in Chapter 2, I focus on data extraction and extractors; namely, practices and entities involved in capturing data from the live world (Internet behaviors, people walking on the street, withdrawing money from an ATM, using Waze while driving, and other digital activities that are prone to data collection). These practices (extraction) and entities (extractors) are highly problematic because they extract data from sources that are incomplete, arbitrary, and generally favor Whites and males such as the Internet, which is by far a diverse place (Bender et al., 2021). All this generates important DEI consequences that are quietly supported (if not promoted) by large data companies, yet they are extremely detrimental to end users (consumers, employees).

References

Ajunwa, I. (2023). *The Quantified Worker: Law and Technology in the Modern Workplace*. Cambridge University Press.

Barocas, S., & Selbst, A. D. (2016). Big Data's Disparate Impact. *California Law Review, 104*, 671–732.

Bender, E. M., Gebru, T., McMillan-Major, A., & Shmitchell, S. (2021). On the Dangers of Stochastic Parrots: Can Language Models Be Too Big? *Proceedings of the 2021 ACM Conference on Fairness, Accountability, and Transparency* (pp. 610–623).

Bowker, G. C., & Star, S. L. (2000). *Sorting Things Out: Classification and Its Consequences*. MIT Press.

Bridges, K. (2011). *Reproducing Race: An Ethnography of Pregnancy as a Site of Racialization*. University of California Press.

Bridges, K. (2019). *Critical Race Theory: A Primer*. Foundation Press.

Cobb-Payton, F., & Berki, E. (2019). Countering the Negative Image of Women in Computing. *Communications of the ACM, 62*(5), 56–63.

Courtland, R. (2018). The Bias Detectives: As Machine Learning Infiltrates Society, Scientists Grapple with How to Make Algorithms Fair. *Nature*, 558, 357–360.

Crawford, K. (2021b). Time to Regulate AI That Interprets Human Emotions. *Nature, 592*(8), 167.

Desmond, M. (2023). *Poverty, by America*. Random House Large Print.

Eckert, P. (1989). The Whole Woman: Sex and Gender Differences in Variation. *Language Variation and Change, 1*(3), 245–267.

Eubanks, V. (2018). *Automating Inequality: How High-Tech Tools Profile, Police, and Punish the Poor*. St. Martin's Press.

European Union. (2021). *Europe Fit for the Digital Age: Commission Proposes New Rules and Actions for Excellence and Trust in Artificial Intelligence*. https://ec.europa.eu/commission/presscorner/detail/en/ip_21_1682

Galbraith, J. R. (1977). Organization Design: An Information Processing View. *Organizational Effectiveness Center and School, 21*(8), 21–26.

Gelman, A., Fagan, J., & Kiss, A. (2007). An Analysis of the New York City Police Department's "Stop-and-Frisk" Policy in the Context of Claims of Racial Bias. *Journal of the American Statistical Association, 102*(479), 813–823.

Howard, A., & Borenstein, J. (2018). The Ugly Truth About Ourselves and Our Robot Creations: The Problem of Bias and Social Inequity. *Science and Engineering Ethics, 24*(5), 1521–1536.

Kalluri, P. (2020). Don't Ask If Artificial Intelligence Is Good or Fair, Ask How It Shifts Power. *Nature, 583*(7815), 169.

Kellogg, K., Valentine, M., & Christin, A. (2020). Algorithms at Work: The New Contested Terrain of Control. *Academy of Management Annals, 14*(1), 366–410.

King, J. T., Yoon, J. S., Rentsch, C. T., Tate, J. P., Park, L. S., Kidwai-Khan, F., Skanderson, M., Hauser, R. G., Jacobson, D. A., & Erdos, J. (2020). Development and Validation of a 30-Day Mortality Index Based on Pre-Existing Medical Administrative Data from 13,323 Covid-19 Patients: The Veterans Health Administration Covid-19 (Vaco) Index. *PLOS One, 15*(11), 1–16.

Lakoff, G. (1987). *Women, Fire, and Dangerous Things: What Categories Reveal About the Mind*. University of Chicago Press.

Luca, M., Kleinberg, J., & Mullainathan, S. (2016). Algorithms Need Managers, Too. *Harvard Business Review, 94*(1), 96–101.

Marabelli, M., Newell, S., & Handunge, V. (2021a). The Lifecycle of Algorithmic Decision-Making Systems: Organizational Choices and Ethical Challenges. *Journal of Strategic Information Systems, 30*, 1–15.

Marabelli, M., Page, X., & Kane, G. C. (2016). Understanding Social Media Use: How Far Will College Student Sampling Take Us? *Academy of Management Meeting* (p. 17704).

Marabelli, M., Vaast, E., & Li, L. (2021b). Preventing Digital Scars of Covid-19. *European Journal of Information Systems, 30*(2), 176–192.

Marabelli, M., Zaza, S., Masiero, S., Li, J., & Chudoba, K. (2023). Diversity, Equity, and Inclusion in the AIS: Challenges and Opportunities of Remote Conferences. *Information Systems Journal, 33*(6), 1370–1395.

Mintzberg, H. (1983). *Structure in Fives: Designing Effective Organizations*. Prentice-Hall, Inc.

Moran-Thomas. (2020). How a Popular Medical Device Encodes Racial Bias. *Boston Review*. https://www.bostonreview.net/articles/amy-moran-thomas-pulse-oximeter/. Accessed on January 8, 2022.

Oswick, C., & Noon, M. (2014). Discourses of Diversity, Equality and Inclusion: Trenchant Formulations or Transient Fashions? *British Journal of Management, 25*(1), 23–39.

Reardon, S. (2023). Alzheimer's Drug Trials Plagued by Lack of Racial Diversity. *Nature*. Published on August 2, 2023. https://www.nature.com/articles/d41586-023-02464-1. Accessed on August 4, 2023.

Rothstein, R. (2017). *The Color of Law: A Forgotten History of How Our Government Segregated America*. Liveright Publishing.

Sánchez-Monedero, J., Dencik, L., & Edwards, L. (2020). What Does It Mean To 'Solve' the Problem of Discrimination in Hiring? Social, Technical and Legal Perspectives from the UK on Automated Hiring Systems. *Proceedings of the 2020 Conference on Fairness, Accountability, and Transparency* (pp. 458–468).

Tarafdar, M., Page, X., & Marabelli, M. (2022). Algorithms as Co-Workers: Human Algorithm Role Interactions in Algorithmic Work. *Information Systems Journal, 33*(2), 232–267.

Taylor, J. R. (2003). *Linguistic Categorization* (3rd ed.). Oxford University Press.

Widder, D. G., & Nafus, D. (2023). Dislocated Accountabilities in the "AI Supply Chain": Modularity and Developers' Notions of Responsibility. *Big Data & Society, 10*(1), 1–12.

Yi, S. K. M., Steyvers, M., Lee, M. D., & Dry, M. J. (2012). The Wisdom of the Crowd in Combinatorial Problems. *Cognitive Science*, 36(3), 452–470.

Wilkerson, I. (2023). *Caste: The Origins of Our Discontents*. Random House.

CHAPTER 2

Data Extractions and Extractors

In this chapter, I discuss practices and technologies used to extract data from what McAfee and Brynjolfsson (2012), in their seminal paper on big data, call "walking data generators." I focus on data extractions (namely, practices that focus on maximizing the collection, storage, and processing of so-called "big data") and data extractors (the technologies used to perform such tasks (extractions)). Nowadays, companies are equipped with very sophisticated technologies that can mine the minutiae of everyone's life. As this book focuses on classifications (and about the importance—and the perils—of classifying), I suggest a classification of extraction practices and associated technologies to systematically analyze them and identify how they relate (positively or negatively) to ethical and DEI issues—we need to realize how such entities (practices and technologies) relate (positively or negatively) to ethical and DEI issues. Practices can be arbitrary, opportunistic, hidden, and deceptive. In fact, while many people have complained about data inaccuracy and risks associated with making decisions with incomplete data in the past, we can now collect extremely accurate data. But this accuracy enables companies to discriminate across populations. For instance, dynamic pricing, a well-known practice that companies enact, refers to a way to change (often in real time) the price of goods and services based on demand/supply. However, the practice often raises ethical concerns because it can

M. Marabelli, *AI, Ethics, and Discrimination in Business*, Palgrave Studies in Equity, Diversity, Inclusion, and Indigenization in Business,
https://doi.org/10.1007/978-3-031-53919-0_2

disadvantage consumers in general and minorities in particular (Nunan & Di Domenico, 2022). Dynamic pricing applies to online sales (e.g., on Amazon) and services (e.g., Uber). When used unethically, dynamic pricing relies on IP addresses and other data to geolocate individuals. These technologies figure out an online customer's zip code (via geolocation) and offer the customer a price that is in line with the spending expectation for a specific area. For instance, in most large cites, there are more and less wealthy areas. I live in Boston and I know that algorithms that manage online dynamic pricing would consider my zip code (Dorchester) as one where people can pay less than others who live in areas such as Back Bay or the South End. At first, one could see practice as a fair way to sell products (of note, Uber does the same when charging riders by areas and neighborhoods). For instance, one could see it as "progressive pricing" that mitigates the effects of Massachusetts' sales taxes (that apply to everyone regardless of where they live, which can be associated with how much money they make). But a more in-depth look at this practice reveals that (1) it clearly attacks people's privacy because they did not receive the option to be charged with respect to their income when accessing a seller's website or service (i.e., there was never a "user signoff," something that represents a principle of consent to collect and utilize personal data) and (2) it implicitly relies on old-school beliefs that American housing is still segregated based on income, which is often a proxy for race. While it is true that zip codes may characterize income and race, the fact that an algorithm assumes so for all residents and, therefore, suggests prices based on the assumption that they can afford it (or not) is deceptive and unfair. It particularly disadvantages low-income individuals who *now* live in areas (zip codes) where most residents are wealthy (fortunately, in cities such as Boston, segregation still appears but not commonly so). It also perpetuates historical prejudices associated with projects and so-called *ghetto* neighborhoods that stem from US redlining policies that were legal during most of the twentieth century and that continue to happen illegally today.

Here, the example should be decoupled into (1) practices (applying dynamic pricing to online sales) and (2) technologies (algorithmic systems that can detect IP addresses and tweak sales prices depending on who connects and *from where*). In the reminder of the chapter, I discuss these practices and technologies more broadly in various contexts of our everyday life with a particular focus on data extractions and extractors.

Sensors constitute the first relevant technology that enables extraction practices.

Sensor Technologies

While sensors have become ubiquitous today, fellow IS scholars Lyytinen and Yoo (2002) pioneered sensors as a concept in acknowledging the pervasive and invasive nature of computing, which began to permeate our lives two decades ago. In 2002, we spoke about people carrying laptops everywhere while beginning to rely on cell phones at all times, which meant that our whereabouts could be tracked 24/7. But the general public had very little knowledge of the privacy implications stemming from constantly communicating with telecommunication towers. Major telecommunication companies were aware of these surveillance capabilities but didn't really understand the value of knowing all this information from their subscribers at that time. In addition, the limited storage capabilities represented constraints to constant monitoring but, more importantly, to mining geolocational data in large amounts. In other words, the technology was there (phones connecting to towers) and the end users were already hooked (dependency on cell phones), but the missing piece was the capability to handle big data. Big data does not simply refer to large data or to Gartner's four Vs (velocity, variety, volume, and veracity) (Beyer & Laney, 2012) but rather to the ability to store, process, and analyze such data. And these computing capabilities were simply not there in the early 2000s. That's why, when we think of big data, we should not refer to data availability but think of whether we can use what we have. Twenty years ago, we had all the four Vs of big data lined up but could not turn their potential into action. Now we can.

But what are sensor technologies? I provide the following working definition for sensor technologies: sensors are physical and digital artifacts that collect data (which includes audio/video footage, geolocation, and biometric data) at the individual and group level often without people's awareness. In this chapter, I unpack this definition and explain sensors' characteristics. Note that this definition differs slightly to one that I have previously offered in a paper on the ethics of algorithmic decision-making systems that has become seminal in the area (Newell & Marabelli, 2015). In contrast to that paper, I consider sensor technology here as a physical or digital artifact (cameras, biometric systems such as a smartwatch, etc.) and its design as important as per the example I discuss later in this chapter with respect to Apple Watch versus Fitbit.

First, we generally think of sensors as physical artifacts, that is, devices with the ability to capture real-world data (a heartbeat, footage, etc.). But sensors might also be digital, such as spyware software (e.g., keyboard loggers). These sensors are embedded in computers and "sense" and report one's activities to third parties such as a hacker. Although "pure" digital sensors might represent a marginal aspect of sensors in general, I think it is worth acknowledging it in the definition.

Second, sensors collect data. That's their job. However, the processing of this data is something that is generally delegated to other systems. For instance, a sensor in an operating room that monitors a patient's blood pressure is made of an automated, non-invasive oscillometric blood pressure cuff that sends the readings to a computer that sets off an alarm if the blood pressure is out of range. But the sensor itself does not set off the alarm. This is the task undertaken by computer software that receives the sensor data. Thus, while sensors have been around for decades, they have only recently become prominent for two key reasons: one concerns business analytics, while the other concerns AI.

With respect to business analytics, with the advent of the big data era, computers can now store myriad data and process it quickly to generate meaningful (or disturbing) outcomes. For instance, massive efforts to collect sensor data about how people drive via onboard devices (such as the one that the car insurance company Progressive[1] provides) gave companies the possibility to store and process these data, cross-reference them with car accidents, and come up with new standards to profile customers as more or less risky as I have explained previously in a study that I conducted with fellow colleagues and friends Sean Hansen, Sue Newell, and Chiara Frigerio (Marabelli et al., 2017). The goal of this monitoring was to offer insurance at a discounted rate to people who were "good" drivers. Insurance companies also sent a driving report to all drivers and marketed the initiative as one that could improve street safety by educating drivers when, in fact, they were collecting data to better profile their customers and assess risks more accurately (with respect to the "type" of customer as they have personal data stored for each insured concerning age, gender, etc.).

[1] https://www.progressive.com/answers/telematics-devices-car-insurance/.

The study I mention above (Marabelli et al., 2017) made my co-authors and I realize the importance of computers' computation capabilities to fully exploit the potential of sensors. The two insurance companies we studied (Progressive in the US and Generali in Italy) could collect driving data such as how often drivers hit the brakes brutally, how often they drove at night, how often they sped up suddenly, and so on. The insurance companies compared these data to databases with probable causes of car accidents. For a period, GPS data was included in the monitoring, mainly to verify whether drivers were speeding. Otherwise, it would be nearly impossible to figure whether a 50 miles/hour speed was within the permitted speed (not knowing whether the driver was on a highway or not). Yet, not knowing current traffic makes it hard to assess whether a certain speed is reckless. For instance, 40 miles/hour on a highway could seem ok (or dangerously low), but it becomes reckless if the highway is packed and someone drives at that speed in the emergency lane or by zigzagging between cars. In addition, it is difficult to keep posted speed databases up to date (e.g., due to roadwork). In sum, it is hard to infer whether someone is a "good" or "bad" driver by using data from an onboard device, even if a GPS is used. Finally, there are privacy issues. A study participant told us that one insurance company was dealing with lawsuits from unfaithful partners being spotted by the GPS in areas they were not supposed to be (note that the report on driving styles (with associated locations in some cases) were sent in the mail to the relevant insured people).

Most importantly, the use of onboard devices by car insurance companies, as we document in our paper, is, in my opinion, highly discriminatory. In fact, regardless of whether GPS location is used, onboard devices capture people's driving style and collect personal behavioral data. Because insurance companies offer discounts to good drivers, wealthy people who can afford to pay the full premium can afford to decline the offer to install an onboard device on their cars. People facing financial hardship, who would otherwise not been able to afford the full premium, need to comply, be monitored, and exchange their privacy with money. In defense of these car insurance companies, it is fair to mention that they also offered an emergency service by which, if a car stopped suddenly (with a deceleration force that exceeded that possible by hitting the brakes), a 911 call would be placed right away with the GPS details on the car's location (assuming a collision occurred). Nevertheless, a fact remains that privacy—in this context—becomes a luxury and a privilege up for sale.

Why sensors have become prominent still relates to storage/process capabilities but departs from business analytics and focused more on expert systems—systems that emulate the decision-making ability of a human expert (Jackson, 1986). The incredible development of AI systems in very recent times has opened up astonishing possibilities to draw conclusions and make predictions based on sensor data. For instance, AI can analyze health biometrics and predict the risk that a patient will face life-threatening disease. A recent study on managing sepsis in hospitals (Sendak et al., 2020) demonstrated that an AI system (Sepsis Watch[2]) could support nurses and doctors in successfully detecting early stages of sepsis. In fact, sepsis, if detected at its outset, is often treatable. Otherwise, it can be fatal. In America, each year, according to the Centers for Disease Control and Prevention (CDC), at least 1.7 million adults in the US develop sepsis, and nearly 270,000 die as a result.[3] While it is important to focus on implementation issues—in fact, in the case of Sepsis Watch studied by Sendak and colleagues, nurses and doctors were involved in developing the AI systems—AI can clearly outperform human judgment. The struggle, according to the study, was that clinicians seldom admitted that an AI was better than them at spotting sepsis symptoms, something that happens frequently when new technologies tend to automate people's jobs in addition to people often not accepting such technology due to the fear that machines will replace them.

But, more importantly, at least due to the focus of this book, with sensors playing such an important role in feeding systems (whether leading to marketing campaigns to attract drivers seeking car insurance or whether leading to better patient care through AI), problem associated with their fairness arise. Despite the fact that AI systems in healthcare are prone to errors as I discuss later in this book, here the "unfair" aspect of all this is that AI-equipped hospitals are generally private hospitals (at least in the US) that require insurance coverage—something that most poor Americans don't have.

Audio/video footage, in a sensor context, equates to surveillance. Examples proliferate. Amazon uses cameras in their warehouses to make

[2] https://dihi.org/project/sepsiswatch/.

[3] https://nigms.nih.gov/education/fact-sheets/Pages/sepsis.aspx.

sure that their employees act as robots (Faraj et al., 2018). Orwellian control and punishment occur if employees don't comply. The management human resources (HR) literature defines this practice as "people analytics" (Gal et al., 2020; Tursunbayeva et al., 2018). People analytics is not just about recording employee behaviors. It is much more. People analytics concerns management's ability to collect data from zoom meetings, emails sent to customers, and performance indicators (e.g., on sales and purchase made). What is key in people analytics is the fact that algorithmic systems analyze user (i.e., employee) data on their work and generate dashboards where employee performance is scrutinized via automated systems. People analytics originated in the late 1950s in the US when Simulmatic Corp. (mashup of the terms simulation and automation) was established in New York City with the goal to use analytics systems to understand preferences of customers and voters. The system was then widely used in the 1960 US elections that led to the presidency of John F. Kennedy (Lepore, 2020). Simulmatic is believed to be the first people analytics initiative to be proven effective and represents the "quantitative turn in the study of man" (Lepore, 2020, p. 37)—an effort to turn emotions, sentiments, feelings, and perceptions into numbers aimed at predicting people's behavior. Nowadays, the term people analytics refers to workforce analytics, HR metrics, talent analytics, human capital analytics, and so on (cf. Huselid, 2018) and describes organizational practices aimed at collecting and analyzing operational data generated by the workforce. This data is captured via sensor technologies, enterprise systems, surveillance cameras, and so on but also via directly querying workers (surveys, self-assessments) and is then made available to management/executives for decision-making purposes. Research to date has pointed out that people analytics may lead to uncomfortable work conditions (due to perceived surveillance) with employees attempting to game the system (when they don't leave the company). For instance, Lee et al. (2015) reported how Uber drivers resisted control by turning off their driver mode when in bad neighborhoods, staying in residential areas to avoid bar patrons, and frequently logging off to avoid long trips. Similarly, Curchod et al. (2019) showed how workers resist algorithmic control by personally negotiating with clients in order to bypass or alter algorithmic ratings in online marketplaces. And Jhaver et al. (2018), in studying Airbnb hosts, illustrated how perceived lack of control and uncertainty over how algorithmic evaluation works can create anxiety.

The point here is that people analytics is not just work surveillance (and discomfort for employees being the victims). People analytics is an attempt, in bright light, to discriminate against non-unionized workers and task workers more generally. In fact, besides very few cases, people analytics is deployed in what Henry Mintzberg, in his seminal book *Structure in Fives*, called machine bureaucracies (Mintzberg, 1980). These are generally large organizations that resemble, from the past, automotive companies such as General Motors and Ford (and their production lines) and, more generally, manufacturing companies with a typical pyramid-shaped organizational chart, with very people at the top, line managers, and 90% of employers who do routine tasks.

Now, the quintessence of these processes is Amazon as I note above. Uber as well, though, at Uber, drivers do not build cars (they drive cars), but the type of (people analytics) surveillance is very similar. Drivers are heavily controlled on what they do rather than on what are the outputs they generate. The populations of the two companies are different. At Amazon, we are not talking about gig economy workers, while at Uber we are. But they are the same, powerless people who are discriminated against because of the lack of protection given by current laws and regulations (here I am mainly referring to the US, but similar examples are not hard to find in most economies). For instance, a 2023 *Wired* investigation on Amazon warehouses[4] reported that first aid clinics in Amazon sometimes directed injured employees to get back on the job instead of seeking professional medical attention to prevent injuries from being documented in hospitals' records. Jordan Barab, a former deputy assistant secretary at the agency for Occupational Safety and Health Administration (OSHA) under the US Department of Labor,[5] posted in a workplace safety newsletter[6] that "What some companies are doing, and I think Amazon is one of them, is using their own clinics to 'treat people' and send them right back to the job, so that their injury doesn't have to be recordable." This example uncovers how the use of sensors in organizations is not neutral. Video-recorded footage of Amazon workers going to the bathroom during work hours is used against them for punishment.

[4] https://www.wired.com/story/amazons-first-aid-clinics-push-injured-employees-to-keep-working/.

[5] https://www.osha.gov/incident-investigation.

[6] https://jordanbarab.com/confinedspace/.

But the same footage does not document workplace incidents, which are then dismissed by Amazon's internal clinics.

Uber is also a good example of sensors (and surveillance) being made of physical but also digital artifacts. As I recently reported in a study with Mareike Mohlmann, Assistant Professor at Bentley University, and Carolina Salge, Assistant Professor at the University of Georgia (Mohlmann et al., 2023), Uber drivers are surveilled by sensors through the Uber app.[7] It doesn't record voices or videos of drivers carrying passengers, yet it records the minutiae of the interactions between the drivers and the customers (riders) and generates performance assessments based on criteria the drivers do not know. This is an example of what I call *ambiguity by design*, which I define as a specific AI design aimed at making a system a blackbox to whoever uses it. For businesses, it makes a lot of sense to design blackboxed systems for two main reasons. First, it helps protect intellectual property related to algorithms. If an algorithm is too transparent, it can be copied by competitors. Examples of social media proliferate (e.g., TikTok). The second reason, still legit from a business perspective yet problematic for users, concerns users possibly trying to game the system. For instance, if Uber drivers knew the exact metrics the algorithm uses to make up their performance, they might try to behave in a way that makes the algorithm "happy" regardless of whether they perform their job effectively. This topic of users gaming algorithms is broadly discussed by (Kellogg et al., 2020) even if they do not specifically discuss Uber drivers. But the example of Uber drivers attempting to game the algorithms can be extended to many different contexts, especially related to people analytics used in companies to, for example, monitor salespeople's performance by looking at their behaviors (number of emails or phone calls to customers, number of visits to customers, number of times they had to negotiate a deal, types of interactions with customers, etc.). Here, too, we do not talk about audio/video surveillance. Yet, the pervasive and invasive nature of control systems make them act as digital sensors. Illegal keyloggers (spyware software that records what we type on our computers) is probably the quintessential digital sensor. In fact, the

[7] Other relevant contribution to the literature (and practice) of the dynamics between Uber drivers and the app were recently made by my colleague and friend Mareike Mohlmann, also coauthor of the paper that is cited above. Examples of her contributions are Möhlmann and Henfridsson (2019), Möhlmann (2021), and Möhlmann et al. (2021).

keyboard is not the sensor, but the software managing it is one. Interestingly, with respect to spyware and monitoring at the workplace, some argue this could be a way to train AI and could potentially lead organizations to replace workers who perform repetitive and routine-based tasks. For instance, in a 2023 interview,[8] David H. Autor, Ford Professor of Economics at the Massachusetts Institute of Technology, suggested that AI could be trained by capturing human-machine interactions; to wit, it is the workforce that, more or less knowingly, will help AI improve itself. He pointed out:

> Often people will be in the process of working with a technology, and the technology will be learning from that interaction. I think there are a lot of discretionary white-collar jobs where you're kind of using a mixture of hard information and soft information and trying to make advanced decisions .... People aren't that good at that, machines aren't that good at that, but probably machines can be pretty much as good as people.

But, as with many others, Autor doesn't envision a scenario with AI creating mass replacement of the workforce—while we will most likely witness a shift of jobs and required competences from task to knowledge workers eventually, in the short term, the shift will likely damage poorly educated people who cannot seamlessly switch from repetitive to more conceptual tasks.

One additional problem of digital sensors—still in the Uber context—is that underlying AI systems hardly learn from users as the former are designed to control the latter. In other words, the machine learning (ML) system that powers the Uber app learns how to better mine user data, such as on how to increase the precision with which riders (customers) receive notifications about when a ride arrives or its estimated time of arrival. However, these systems are not designed to be "user-friendly" to use a term from the user experience literature. With Monideepa Tarafdar, the Charles J. Dockendorff Endowed Professor and PhD Coordinator in Information Systems at the University of Massachusetts Amherst, and Xinru Page, Associate Professor at Brigham Young University, I have studied interactions happening between humans and algorithms in the Uber driver context. We theorized that feedback happens through two pathways from the human to the algorithm. The first one is based on

[8] https://www.wired.com/story/corporate-surveillance-train-ai/.

humans' task actions that are automatically recorded through the Uber app. They form the inputs that are given to ML and are used to calculate subsequent outputs regarding future tasks, performance, and incentives. This is illustrated by the computational learning feedback loop through which the actions recorded by the app are fed back into the algorithm. The second pathway relates to humans' cognitive reactions, e.g., the frustration, dissatisfaction, anxiety, and reasoning associated with them. What is problematic is that the second pathway recorded by the app is not communicated back to the algorithm for future computations. For example, drivers do not (and cannot) tell the algorithm that they did not understand why their passenger rate decreased suddenly. Algorithms do not learn from human's emotional reactions. We call this *broken loop learning*. This is highly problematic and showcases the opportunistic approach that companies (not just Uber of course) take when designing algorithms that manage the workforce. With Uber, this issue is exacerbated by the fact that the drivers are gig economy workers rather than employees and, therefore, have less protection. For instance, they cannot join a union and can hardly protest. In defense of Uber, however, it is worth noting that several Uber drivers prefer to be contractors rather than employees because this status gives them more flexibility with respect to their work schedule.

An additional issue concerning Uber's many ways to control their drivers is about a recent evolution of the AI systems embedded in the Uber app that manages rides. Since 2021, Uber has used Microsoft Face API software[9] in its app to verify drivers' identification. Drivers are asked to submit new photos on a regular basis. In some instances, the software has failed to accurately recognize people with darker skin tones. The fact that facial recognition systems are biased against people with dark skin is well-known. Kate Crawford, Research Professor at the University of South California Annenberg and a Senior Principal Researcher at Microsoft Research Lab in New York City, broadly discusses this issue in her book *Atlas of AI* (Crawford, 2021a). Here, privacy issues emerge once again (see discussion on little data later in this chapter) but so too do discriminatory practices put in place by these apps. In fact, being recognized by the app via facial recognition is a requirement to begin working. This might prevent people with dark skin from making money,

[9] https://azure.microsoft.com/en-us/free/ai-services/.

paying their bills, and overall having the same chances as White people to avoid being miserable in our society. Interestingly, and related to my previous point of people trying to game AI system, a Google search (performed on July 13, 2023, from my Boston home) showcased the following (Fig. 2.1).

Google suggested very interesting search options (or "predictions" as it likes to call them). The first thing worth noting is that photo verification is problematic (not working, failed). The second thing to consider is that these issues are broadly discussed in online forums (e.g., on Reddit). The third thing concerns gaming the system (how to bypass photo verification). The conclusion is that what Uber does is highly problematic and potentially racist (because of the software it employs) and that blaming drivers because they try to find shortcuts to access a job, an income, and a decent life is at least questionable. This very simple experiment (Google search) showcases the struggles gig economy workers experience when surveilled. The issue here is twofold. On the one hand, gig workers feel extremely uncomfortable when scrutinized while working (focus on processes rather than outcomes, which is, however, typical of machine bureaucracies). This type of control resembles century-old principles of scientific management that Fredrick Taylor (1911) created. He elaborated a technique to precisely measure workers' tasks (literally with a clock in his hand) to identify areas of improvement (i.e., when workers

Fig. 2.1 Google search related to Uber photo verification

could do things faster or better). But many after Taylor have demonstrated that these practices are detrimental to workplace wellbeing, such as Barnard (1938), who was the first to discuss informal aspects of organizations, comprising social networks and interactions and where leadership replaced power and hierarchy. In current times, especially with the advent of the gig economy, we are witnessing a revival of Tayloristic principles, one example being crowdworkers who are paid (or underpaid) for specific tasks measured using time, number of micro-tasks, or similar quantitative outputs. What makes it exploitation is the fact that crowdworkers generally perform such work because they can't find anything better to make ends meet. What is more, crowdworkers are often underqualified for the jobs they undertake. Yet, they are used, for instance, to train AI systems—something that I discuss in Chapter 3.

While I do not dive into discussing the role of geolocation systems with respect to sensors—Uber is already a good example of how sensors mine workers' whereabouts—three key aspects of my definition of sensors still need to be discussed: the way sensors collect data at the individual and group level, people's awareness when they are being monitored by sensors, and biometrics. Note that awareness and biometrics are highly connected. Some people argue that sensors are more targeted to individuals than groups and, thereby, less discriminatory in contrast with systems that collect group-level data and infer generalizations. This parallel can be found in the literature discussing big and little (or smart) data (cf. Loebbecke & Picot, 2015; Newell & Marabelli, 2015). With big data, companies generally rely on statistics when trying to sample populations, such as to sell products to a specific group of users.

But sensors are more tied to little data. Little data does not rely on statistics. Instead, it offers individuals deals based on their specific behaviors. Going back to my study on the automotive insurance industry, before the advent of sensors, these companies relied on statistics to make up the premium on individuals (the "traditional" job of actuaries). For instance, data showcased that young single men were more prone to get into car accidents than middle-aged women with kids. Therefore, companies charged men more than women, especially based on the age bucket they would fall in. But this is very discriminatory (as most generalizations are). In Europe, this practice became illegal. For instance, consider diligent and prudent 18-year-old drivers who pay out of pocket to attend college and who must pay more to be being able to drive a car only because some of their peers are not good drivers. Discriminating against individuals for statistical reasons, or generalizations, is something that has permeated our lives for centuries. It leads to stereotyping populations. Italians are corrupt, Irish-Americans get into fights in bars, uneducated people are less smart than people with a college degree, Black Americans commit more crimes than Whites. Besides being nasty and humiliating, these stereotypes are *classifications* leading to exclusion practices, marginalization, and unfair treatment of human beings. As I pointed out almost a decade ago (Newell & Marabelli, 2015), big and little data represent a tradeoff. Big data leads to discrimination. Little data leads to privacy issues. These are two key features of *datafication* (cf. Marabelli & Markus, 2017).

Group-level sensor data is a rather overlooked yet key characteristic of sensors. It surfaced with the COVID-19 pandemic with digital contact tracing systems. These systems (developed as smartphone apps) were developed to spot smartphones equipped with the app in close contact (via Bluetooth, wireless, or GPS). If John and Adam had their smartphones with them, the same digital contact tracing on their phones, and came into close contact (say 10 feet) when shopping on a given morning at the same grocery story, then the app could contact one of them to take a COVID test if the other found out he had contracted COVID. It is clear that digital contact tracing systems can work to notify people about potentially being infected insofar as they own a smartphone, have downloaded the necessary app, and have activated it properly. Digital contact tracing is the quintessence of contemporary sensor technologies. But they come with many ethical and DEI concerns. Digital contact tracing is also a good example of people's distrust in institutions. In fact, in countries where the technology was not mandatory (most democracies), very few people adopted it (an interesting study by Seto et al. (2021) provides more details and support to my point). But let's focus on problems with digital contact tracing systems from an ethical and DEI standpoint.

First, digital contact tracing poses privacy issues. The geolocation systems on which tracing apps rely on reveal people's whereabouts at all times (if the apps activate as they should). A 2021 Brooking study[10] revealed that 45% of Black Americans supported the technology versus 61% of White Americans. While this is only a slice of the US population,

[10] https://www.brookings.edu/articles/contact-tracing-apps-face-serious-adoption-obstacles/.

it also reflects the lack of confidence that Black Americans have in the government because of surveillance. A 2022 report[11] concerning stop-and-frisk in New York City filed by the American Civil Liberty Unions noted:

> The NYPD has conducted millions of stop-and-frisks in New York City over the last two decades. The majority of those stopped are people of color, and a vastly disproportionate number are Black. There is no evidence that ramping up stops makes New Yorkers safer. But we do know that many of these stops have been unlawful and that some have led to violent police misconduct.

Frisking in America involves searching citizens for illegal possession of weapons. Stop-and-frisk happens when cars are stopped for this purpose. To justify the stop "and in justifying the particular intrusion, the police officer must be able to point to specific and articulable facts which, taken together with rational inferences from those facts, reasonably warrant that intrusion" (cf. the Terry v. Ohio US Supreme Court deliberation, in 1968[12]). I think that this example provides a good rationale for explaining why the majority of Black Americans would refrain from using digital contact tracing apps. Privacy in the US is a work in progress, especially if we think of Canada,[13] the EU (cf. GDPR[14]), or even China where a comprehensive (even if not perfect) set of regulations was put in place in 2021 (Lewis, 2023).

Second, Black communities in America may see digital contact tracing with skepticism because of the longstanding "bad" relationships between these communities and healthcare. Why would Black Americans believe that they would benefit by downloading an app on their smartphone that can track all their whereabouts given that the US healthcare system has been historically (and currently) so cruel to them?

Third, digital contact tracing requires one to have a smartphone. According to a 2021 PEW study,[15] roughly 85% of Americans have a smartphone, which means that 15 million Americans would potentially

[11] https://www.nyclu.org/en/stop-and-frisk-data.

[12] https://supreme.justia.com/cases/federal/us/392/1/.

[13] https://laws-lois.justice.gc.ca/eng/acts/p-21/FullText.html.

[14] https://gdpr.eu.

[15] https://www.pewresearch.org/internet/fact-sheet/mobile/.

not be able to access a key resource to stay healthy. Here, it is important to note that some might say that 85% of Americans owning a smartphone would be a good reason to focus on digital contact tracing as a technology to protect citizens from the next pandemic. The argument has two ramifications, one technical and one societal. The technical one concerns the efficiency of digital contact tracing. The more people use it, the better it works. Obvious but worth stressing. An economist would say digital contact tracing benefits from network externalities. Research from the University of Oxford suggests 60% uptake among a population as the minimum threshold for digital contact tracing to work effectively.[16] Therefore, the above statistic for how many Americans have a smartphone would support (at least in theory) a push to use such technologies in the US. That doesn't mean that we should not promote and use this sensor technology in the future. But an ethical approach should be inclusive rather than exclusive. Because we have enough smartphone users in America that might be persuaded (or coerced) to use digital contact tracing during a pandemic, how do we take care of the remaining "minority"? It is always hard to think about minorities. But while focusing on the "majority" might lead to better outcomes (for most), it automatically leaves behind the poor, marginalized populations, and historically underprivileged communities. Besides that, addressing the poor trust that many Black Americans have in healthcare and government might widen the percentage of users who would actually use digital contact tracing.

I'd like to stress this focus on the majority because decision-makers generally use this "principle" to make highly discriminating choices. In a recent paper that I wrote with colleagues and friends Sam Zaza, Silvia Masiero, Lydia Li, and Kathy Chudoba (Marabelli et al., 2023), we broadly discussed DEI issues associated with running conferences and argued that hybrid conferences should be the right format because they are more inclusive. Lessons from COVID-19 have shed light on how more inclusive conferences were when they were online. For instance, online conferenced allowed many people from the Global South to access, for the first time, major IS and management conferences (based on a survey we conducted for the paper). But now, sadly, most conferences have reverted to in-person formats. As a current and former conference organizer, I have good knowledge of how these decisions are made.

[16] https://www.research.ox.ac.uk/article/2020-04-16-digital-contact-tracing-can-slow-or-even-stop-coronavirus-transmission-and-ease-us-out-of-lockdown.

Besides reviewing costs of running conference (hybrid conferences are the most expensive to organize because one needs to pay for a physical venue in addition to hybrid technologies that typically only venues in "good" districts in large and expensive cities can offer), the conference organizers surveyed participants. And surprise, the majority of the respondents favored in-person conferences. Why? Because in-person conferences are more fun, people can visit cities and travel, social events promote networking, and so on. But the problem is that the majority (the majority of academics) comes from wealthy countries and universities who pay for their travel and lodging. It is here that we need to listen to the minority, people whose voice is hardly heard, because they seldom participate in conferences and, therefore, may play a less relevant role in academia. And if we keep making decisions based on what the majority likes, we will likely perpetuate exclusion practices.

Going back to the digital contact tracing example, if most Americans are happy with it, that doesn't mean that all government resources should be devoted to developing the technologies that enable it. It is definitely worth working on digital contact tracing, but some resources need to be spent for people who do not have access to devices that only the majority owns. Achieving DEI requires a deontological approach to decision-making, by which ethical principles should be followed regardless of the outcomes; if a hybrid conference can serve more people (even if with a lower-quality "conference experience" because in-person attendees are somehow annoyed by big screens in a room with online people), we should go for it. This is in contrast with the dominant, capitalistic, and utilitarian approach that goes under the consequentialism umbrella according to which what is relevant is the outcome. If more people are happy with one outcome (in-person conferences), that should be the decision made. Consequentialism emphasizes maximizing overall happiness, wellbeing, or utility as the fundamental principle for ethical decision-making. Here, I argue that consequentialism might be detrimental to DEI in general terms and with respect to how we develop and roll out technologies, especially if we talk about expert systems such as AI.

People's awareness of sensors is another important issue. The general public likely realized how surveilled people can become totally unaware of the fact they are being observed happened with the first edition of the Big Brother, aired by CBS since 2000 (note that the Big Brother was actually created by John de Mol, a Dutch producer, and was aired in the Netherlands in 1997, but it was the American version that led to the widespread diffusion of the format). That surveilled people can forget about cameras and microphones recording their everyday (and overnight) actions is something that George Orwell understood several decades earlier in 1949 when he published his book *Nineteen Eighty-Four* (from which the term "Big Brother" was borrowed). Forgetting about being surveilled is what probably happens to inmates in what Matt Burgess, in a 2023 *Wired* article, called "smart prisons"—detention facilities equipped with myriad sensors (he mentions a jail in Fulton County, Atlanta, Georgia, as an example). These sensors control inmates 24/7 in a way that is everything but visible to them. For instance, wall sensors can capture inmates' movements, the proximity of two or more inmates, and even heart rate details, such as historical data of their biometrics, using software produced by Talitrix[17] called "Inside the Wall." Another example concerns how someone can (illegally) use GPS-equipped devices such as Apple AirTags (small devices that one can attach to keys or put in their bag) to spy, abuse, harass, and stalk someone else.

Referring to previous people analytics examples, positive consequences concern managers' ability to spontaneously capture behavioral data (how people do things) about employees. This should lead to more objective performance assessments. Negative consequences concern invasive and/or sneaky ways to monitor individuals in society. Invasive (yet totally legit) ways to monitor individuals relate to wearable devices collecting biometric data, most notable examples being Apple Watch and Fitbit (but they also include wristbands, hearing aids, and smart clothing such as "e-pants"[18]). These sensor technologies are purchased by the wearer (so here there's perfect awareness) but, with use, people sort of forget that they are continuously feeding these devices with their own personal data, such as their as heartbeat, blood pressure, body temperature, and so on. The problem here is that, depending on the design of these devices, the data collected is more or less prone to be shared with others (individuals and companies), which has consequences. The design of sensors such as the Apple Watch and Fitbit is dramatically different, which has implications for privacy and discrimination. The Apple Watch is designed so that biometric data is sent from the watch to one's phone. The phone then showcases charts, stats, and health suggestions (you should sleep more,

[17] https://www.talitrix.com/inside-the-walls/.

[18] https://theintercept.com/2023/09/02/smart-epants-wearable-technology/.

your heart rate during the 1,500 steps was too fast, etc.). The data stays with the user assuming that Apple is faithful to its promises to not download user data (but Apple has a good record in this regard). Fitbit works differently: it connects with an online cloud, which processes the data and sends reports to the user's phone. This latter example has two main implications: first, the company owns the data (or at least has free access to it). Second, should a security breach affect Fitbit, all health data from its user would be in jeopardy.

But what are the "real" consequences for end users? Put differently, why is it so dangerous that health data are shared with third parties? Let's first examine how personal health data can be *extracted* in more or less subtle ways. As I wrote recently (Marabelli et al., 2021), Google owns Fitbit, and, by collecting data on even a few users who shared their data, the company can predict health outcomes for much larger groups with similar characteristics and, therefore, target a much wider array of consumers via its platform. In this way, it can generate invasive advertisements and potentially create discrimination. For instance, genetic information is considered sensitive information that can lead to job and healthcare discrimination. Yet, only 2% of a country's population needs to have done a DNA test to identify nearly everyone else (Erlich et al., 2018). Learning that (statistically) young people are more prone to have accidents or that people with kids are more prudent drivers shouldn't allow car insurance companies to charge them different fees.

The genetic industry has recently taken a sharp right turn when law enforcement started to collect such data (even if, for now, on a voluntary basis). In September 2023, state police in Connecticut held what they called a "DNA drive" reportedly to, among other reasons, identify found human remains. Family members of missing people were invited to submit their DNA in the hope to find positive matches with bodies. Yet, as a 2023 *Wired* article suggested:

> Genetic genealogy isn't used by law enforcement only to identify missing persons and human remains. It is also widely used to identify suspects in investigations. Even if a suspect has not submitted their own genetic profile to a consumer site, investigators can infer biological relationships based on how much their DNA recovered at a crime scene matches that of other

users. Police use that information, along with public records, to build out a suspect's family tree and narrow down their identity.[19]

This adds to the very loose regulations on how DNA data should be handled by commercial companies such as 23andMe[20] or AncestryDNA[21] with data being regularly acquired by law enforcement to pursue investigations. While it might be alright, in principle, to use available technologies to prosecute, the underlying issue is threefold. First, privacy issues arise since DNA evidence can identify people who did not agree to be in a database (e.g., someone's relatives). Second, genetic data is shared with third parties in unregulated ways, which makes DNA companies such as 23andMe and AncestryDNA de facto data brokers (something that I discuss later in this chapter). Third, even if one ignored the lack of regulations about sharing data and assumed that sharing companies would act ethically, no system is 100% secure and safe from cyberattacks. Indeed, simply storing such sensitive (genetic) data creates the potential for data breaches. For instance, in October 2023, 23andMe reported a large data leak with hackers able to guess users' login and password (supposedly with brute force and dictionary attacks). Once the hackers gained access to a group of users, they were able to access data about additional users with a function on the 23andMe platform called DNA Relatives, which allows users to opt in to share their genetic information for others to see on the same platform. The stolen data was soon found on the dark Web for sale and included over one million data points exclusively about Ashkenazi Jews.[22]

In America, the consequences of someone having their health data shared with third parties are particularly problematic because its healthcare system mainly relies on private companies. Just to set the record straight, America, like most countries, has a double system: private and public healthcare. It is important to clarify this because many people—including several (privileged) Americans—believe that public healthcare here doesn't exist. Public healthcare *does* exist, is relatively widespread, and provides decent care to people who can't afford to pay medical bills.

[19] https://www.wired.com/story/dna-drives-help-identify-missing-people-its-A-privacy-nightmare/.

[20] https://www.23andme.com/dna-health-ancestry.

[21] https://www.ancestry.com.

[22] https://www.wired.com/story/23andme-credential-stuffing-data-stolen/.

For more information, I suggest a landmark book on the topic by Khiara Bridges called *Reproducing Race: An Ethnography of Pregnancy as a Site of Racialization* (2011). The problem is rather cultural. Employers provide (or *have* to provide) private healthcare to full-time employees. As a consequence, only few people (typically unemployed and part-time workers) use public safety nets. Also, public healthcare pays less; therefore, it is not uncommon to find more qualified doctors in private healthcare. This is, however, not a rule as I know personally several excellent doctors working in the public system just to help people who can't afford private care.

But all that said, it is a matter of fact that, in America, most people rely on (and pay for) private healthcare. This means that they pay a monthly fee to health maintenance organizations (HMOs) who are the insurance companies that pay the bills to hospitals. Once insured people need care (e.g., on a one-off basis because they fell from their bike and need urgent care or because they need monthly prescriptions for blood pressure), the HMO pays the carer (hospital, pharmacy) and charges the insured people a minimal part of the cost (so-called copay). I purchase two epi-pens every year because I am allergic to peanuts. The market price for these emergency medications is around US$300. But I pay US$20 (my copay), while the rest is on my insurance (the HMO). Here, there are two issues. One is about what is covered and what is not. Generally, emergency procedures are covered almost in full, while plastic surgery is likely to be not covered by any insurance company. The second and more relevant issue concerns preconditions. HMOs want to make money off the people who they insure (why wouldn't they?). Right now, health insurance regulations in America prohibit HMOs from charging individuals higher premiums based on preexisting conditions due to the provisions established by the Affordable Care Act[23] (ACA). Under the ACA, insurance companies must offer coverage to individuals regardless of their health status or preexisting conditions. However, it's important to note that specific details and regulations can evolve over time, and there may be variations in the implementation of these regulations at the state level or in specific insurance policies. For instance, it is possible that new administrations will repel the ACA as almost occurred in recent times. In 2016, the late US Senator

[23] https://www.healthcare.gov/glossary/affordable-care-act/.

John McCain was the one person in the US Senate who prevented it from happening.[24]

Therefore, preexisting conditions and associated insurance coverage is a very fluid matter. It is for this reason that it is important to protect personal health data. What if, today, Fitbit learned that I have hypertension and tomorrow I couldn't purchase health insurance simply because Fitbit sold my data to an HMO or because Fitbit had a security breach and my health data connected to my smartwatch became public? Sensors collecting biometrics deal with very sensitive data, which can determine how much people will be charged for their insurance (or even whether they will be covered by any HMO). Discrimination here does not concern the color of the skin but the health of the body. If you are already ill, you will get more ill because you will be denied care (unless you are wealthy of course). And because you will need urgent care, you will be even more sick because you will have to rely on public care (safety nets) that is less responsive and slower than private care in seeing you, examining you, medicating you, and so on. Sensors, in the form of wearable devices, can lead to all this. Genetic data mining can lead to similar consequences; that is, to discriminating people even before they become ill. For instance, genetic tests that screen women concerning hereditary breast and ovarian cancer are available today. What if HMOs could run these tests on people whose DNA is available and had to make a decision on whom to insure (e.g., between someone with a positive and someone else with a negative test)? Insurance companies (car and healthcare insurance companies) survive—and make a lot of money—when the people they insure are healthy (or drivers don't get into accidents). HMOs are revenue-driven, so it is not hard to imagine what they would choose should they have genetic data in their hands and the legal right to turn down applicants based on such data.

What makes health issues managed by sensors even more problematic, from a privacy standpoint, is that companies such as Apple and Google are not regulated under the Health Insurance Portability and Accountability Act (HIPAA),[25] which Senators Edward Moore Kennedy

[24] https://www.npr.org/2017/07/27/539907467/senate-careens-toward-high-drama-midnight-health-care-vote.

[25] https://www.hhs.gov/hipaa/index.html.

and Nancy Kassebaum Baker proposed in 1996 and the Clinton administration approved. HIPAA regulates the flow of healthcare information and spells out how patient personal data needs to be protected (data thieves) and dealt with (it prevents healthcare providers from sharing health data with third parties, which includes an individual's relatives unless the individual has signed an agreement, i.e., is someone's health proxy). In brief, healthcare companies and organizations (public and private hospitals, HMOs, medical labs, etc.) in America are upheld to very high privacy standards when dealing with medical data. However, high-tech companies that constantly deal with sensitive health data do not fall under the HIPAA's rules because they do not constitute healthcare organizations even though they deal with healthcare data and probably have more such data than most healthcare facilities that need to comply with the HIPAA.

In fact, if you wear a smartwatch that constantly monitors your vitals, who do you think has more health data about you? Your smartwatch or your doctor? Probably, one aspect of these sensor devices that keep them out of the scope of current laws is that they are very well received from the general public who seldom complain about the devices with respect to privacy. People find wearable device enjoyable and helpful to keep fit, trace sleeping (and be aware of sleep deprivation), and even be "happier." For instance, Fitbit currently offers a "mood logging" feature among its stress management tools.[26] Clue and Flo are apps that track women's period and how it relates to their moods. The fact that most people are generally fine with their privacy being breached is illustrated by the countless Facebook data scandals (the most notable example being the 2018 Cambridge Analytica scandal,[27] which involved 87 million users) that have resulted in almost no impact on user attachment to the platform. Even the negative reaction of Facebook's investors (Facebook shares fell more than 24% in the aftermath of the Cambridge Analytica scandal) was temporary, and, one month after the scandal, the public value of the company had completely restored.[28]

But healthcare privacy associated with smartphone use took a sharp right turn on June 23, 2022, when, in the Dobbs v. Jackson Women's

[26] https://www.fitbit.com/global/us/technology/stress.

[27] https://www.nytimes.com/2018/04/04/us/politics/cambridge-analytica-scandal-fallout.html.

[28] https://www.cbsnews.com/news/facebook-stock-price-recovers-all-134-billion-lost-in-after-cambridge-analytica-datascandal/.

Health Organization,[29] the US Supreme Court made a landmark decision that the US Constitution does not confer the right to abortion, a decision that overruled two previous rulings (Roe v. Wade in 1973 and Planned Parenthood vs. Casey in 1992). The Dobbs ruling left the decision to states on whether abortion would be legal and in which terms and with which limitations. This created a situation where abortion became illegal (or had several limitations) in some states. The technology implications concerning sensors are immense. In fact, using location data, various kinds of brokers have the ability to collect data about people's movements, identify potentially pregnant women based on their online behaviors and Web searches, and see whether they try to move out of state to interrupt their pregnancy.[30] Black women seek abortions at higher rates because, among other reasons, they often have poor access to contraceptives and because their pregnancies are more prone to clinical complications (due to poor access to quality care[31]). And Black women live prevalently in Southern conservative states where abortion is banned because of the focus on religious principles and the "pro-life" rather than "pro-choice" approach. In turn, Black women are disproportionally affected by the Dobbs ruling, and sensor technology contributes to it.

In conclusion, sensor technologies are invasive and pervasive and pose several ethical and DEI issues. They are effective data extractors at any time of the day (as per smartwatches when we sleep). These data can be used for purposes that deviate from the initial reason why they were collected. For instance, the Uber app collects estimated time of arrival (ETA) data seemingly to provide a better service to customers, but Uber can use this data to fire drivers if they don't meet the ETA according to the algorithm. In a similar way, data collected by insurance companies can jeopardize people's privacy, which can have significant consequences. From the examples laid out above: (1) in the automotive industry, wealthy people can "buy" their privacy; (2) in healthcare, people who are already sick (with preexisting conditions) risk receiving less care (because they are more sick, which is a paradox). Here the point is that sensors represent an opportunity to progress. But irresponsible use of this technology is

[29] https://supreme.justia.com/cases/federal/us/597/19-1392/.

[30] https://www.cnn.com/2022/07/16/politics/abortion-data-what-matters/index.html.

[31] https://www.washingtonpost.com/politics/2022/06/25/dobbs-roe-black-racism-disparate-maternal-health/.

detrimental to privacy, can lead to discrimination, and, more generally, poses DEI issues.

Web Scraping

One of the first things that I mention to my students in the first class for my course on business analytics at Bentley University where I explore the opportunities and challenges of using algorithmic-based technologies is the need to discern between what is ethical and what is legal. For instance, it is ethical to feed homeless people, yet it might be illegal to offer food to people living on the street. In Newark, New Jersey's largest city, a permit is required to feed homeless people.[32] But it is legal (at least in several US states) to leave a car parked with the engine (and air conditioning) running while grocery shopping for a certain amount of time—something that, to me, appears highly unethical. Where this is not legal, it is hardly enforced.

The Internet is a much messier (and less regulated) place than the street, and it is understandable that things get out of control. This is not necessarily a negative thing. The Arab Spring was widely supported by social media, which allowed people to organize protests, disseminate information about their activities, and raise local and global awareness of ongoing events (Wolfsfeld et al., 2013). However, people often use data available on online platforms in ways that are (in the least) debatable. I do not get into specific clearly unethical and illegal online behaviors here, such as the generation of fake news, harassment, online bullying, threats, stalking, and so on. Instead, I focus on Web scraping, a poorly regulated practice that involves extracting information from the Web in both a manual and, more commonly, an automated fashion with specific software. Problematically, Web scrapers often ignore the difference between *public* and *available* information (or are not trained to do otherwise).

I discuss the ethical/lawful debate in my classes in other ways as well. For instance, I identify a random student and ask whether the student has a social media presence with a pubic photo (e.g., a profile photo on LinkedIn). The answer is always a solid yes. Then I ask: what if you came to my office next week and saw that photo as my computer's background? At this point, the students I have asked usually give me a strange

[32] https://www.nytimes.com/2021/12/14/nyregion/newark-prohibiting-feeding-homeless.html.

look before understanding what I mean. Then they say "that would be creepy." This is the right answer. It would be inappropriate but not necessarily illegal. Just creepy. If we think of this process of appropriating user-generated content in an automated manner, we see how our data is constantly exposed to being mined by companies that want to profile individuals based on their online presence. A typical data extractor is a Web spider that systematically scans webpages to gather information on a specific subject. For instance, recruiting companies often use these techniques to fetch data from people's LinkedIn profiles to generate datasets of people, skills, current jobs, and so on. These companies offer these data to other companies in search for talent. Sales companies use social customer relationship management (CRM) systems to identify potential customers by automatically scraping the Web in search for user contributions (forums such as Quora, Reddit, or social media). Going back to my example of photos one can download and use as they want, the way companies scan and scrape the Web very much violates people's privacy. Data being collected is used for a purpose that is different than the data that was initially uploaded for. For example, students don't post their photo on LinkedIn to give others the possibility to store it on their computer. Similarly, people's professional skills should not be mined by companies as these skills are not always posted with the goal of looking for a job.

Besides privacy issues that I discuss above, Web scraping poses relevant DEI implications. First, not everyone is online for various reasons. If recruiting companies build their database based on Web scraping, they will necessarily leave out some people who might be looking for a job. Generally, people with an online presence aimed at job hunting have or have had the means to do so. In other words, the Internet job market is one for privileged individuals, which excludes, for instance, the elderly, a category that is often discriminated against during hiring processes (even though laws in America prohibit discrimination based on age). Second, data scraped from the Web by recruiters can be shared (or sold) to other companies that make money out of the scraped information. Examples include companies within the data brokerage industry (currently, a virtually unregulated practice in America at present[33]) that collect Web data and match them with other personal data (e.g., from the juridical system).

[33] https://www.wired.com/story/opinion-data-brokers-know-where-you-are-and-want-to-sell-that-intel/.

In this way, they can match (generally public) arrest records with data associated with a person's arrest (such as mug shots, home address, and more) and make it available for sale. But these records include people who were wrongly arrested. In a 2021 *Wired* article, Sarah Esther Lageson reported[34]:

> In April 2018, Adnan (a pseudonym) was wrongfully arrested in Newark, New Jersey, on the basis of an incorrect arrest warrant. A brief period in jail led to the criminal court judge dismissing the incident and moving to have Adnan's arrest record expunged. A few days later, Adnan began receiving mysterious text messages from several "reputation management" companies that promised to help him get his mug shot removed from the internet. A few days after that, he received an email from a company that specializes in "arrest-related content removal." Confused, Adnan contacted the local and state police departments to ensure that his expungement order had been received from the courts. It had, but Adnan's arrest data was already in the hands of data brokers.

This example raises two problems. The first and more general problem involves all citizens and the inability to remove online records even if they don't correspond to the truth, which can prohibit people from gaining new work, insurance, a mortgage, and so on. As I previously explained, algorithms that vet people (e.g., to access loans) are nebulous and are fed, in part, with online data. If one's Facebook friends can jeopardize a mortgage application, imagine the negative effect of an alleged (online) criminal record. The second and more specific problem relates to race. In America, Blacks are more prone to be scrutinized (e.g., the stop-and-frisk example earlier in this section) and, thereby, experience disproportionally higher arrest rates and are more likely to face serious charges and serve longer jail time than other populations (Sommers & Marotta, 2014). Data extractions that rely on automated systems are not neutral. They punish people already struggling in our society and turning over government (federal, state) records to private companies promotes what Shoshana Zuboff, Professor Emerita at Harvard Business School, calls "surveillance

[34] https://www.wired.com/story/criminal-justice-transparency-law-data-brokers/.

capitalism" (Zuboff, 2019) where businesses make money out of individuals' private data. In a recent (2021) *New York Times* opinion article,[35] she stated:

> Facebook is not just any corporation. It reached trillion-dollar status in a single decade by applying the logic of what I call surveillance capitalism—an economic system built on the secret **extraction** and manipulation of human data—to its vision of connecting the entire world. Facebook and other leading surveillance capitalist corporations now control information flows and communication infrastructures across the world.

But the point here is that the targets of data extraction are not just any citizen. Wealthy people can protect their data, such as by paying (other) companies to have their online records cleaned (note that, in Europe, GDPR's *right to be forgotten*[36] is, as the phrasing says, a (free and accessible) right, not a privilege) or by paying a higher car insurance premium because they want to avoid being tracked. In America, the targets are people at the margin who have no control over their data, and they often have no idea that companies trade their data, which leaves them to wonder why they can't access credit, gain health insurance, and/or find work.

In America, data brokerage is a practice that occurs actively between private companies and law enforcement with the latter often purchasing individuals' personal data. Brokerage companies include Acxiom,[37] CoreLogic,[38] and Epsilon.[39] These companies systematically extract user data from the most disparate sources often for marketing purposes and with the idea to resell personal data to businesses. One such business (as an example) is Oracle, which, according to a 2019 *Financial Times* report, works with over 80 brokers.[40] According to a 2023 *Wired* article by op-ed contributor Justin Sherman, law enforcement agencies purchase data from brokers in a systematic fashion. The US Department of Homeland Security (DHS) and its sub-agencies such as Immigration and Customs

[35] https://www.nytimes.com/2021/11/12/opinion/facebook-privacy.html.

[36] https://www.dataprotection.ie/en/individuals/know-your-rights/right-erasure-articles-17-19-gdpr.

[37] https://www.acxiom.com.

[38] https://www.corelogic.com.

[39] https://www.epsilon.com.

[40] https://www.ft.com/content/f1590694-fe68-11e8-aebf-99e208d3e521.

Enforcement (ICE) (which has put children in cages[41]) have purchased cell phone location data on millions of Americans, home address information to support deportations, and home utility data for investigations. This is a typical example of how data extraction is part of a supply chain (involving government agencies) that trades personal data for money—the quintessence of surveillance capitalism as Shoshana Zuboff would say—and discriminates against and takes advantage of people who are not in the condition to hire a lawyer or anyway benefit from government protection (also, because in the case of ICE, they are the institutions). As Sherman argues: "the impact of stockpiling individuals' personal information will fall hardest on the already oppressed or marginalized: the poor, Black and brown communities, Indigenous populations, LGBTQ+ individuals, undocumented immigrants." The good news is that, on July 19, 2023, the US House Juridical Committee passed *The Fourth Amendment is not for Sale Act*[42] with bipartisan support (the Four Amendment concerns "The right of the people to be secure in their persons, houses, papers, and effects, against unreasonable searches and seizures…[43]"). This bill, if turned into law, would stop government agencies from buying data without a warrant (the US Constitution, through the Fourth Amendment, protects people from unreasonable searches and seizures by the government).

One final issue concerning extracting data using Web scraping concerns what is actually on the Web and how the information is used to make decision. The wisdom of the crowd is supposedly always right because it results from many different individuals, especially if we consider fake news and false information as a relatively marginal phenomenon (it is actually NOT a marginal phenomenon but, so far, not enough to bias online content). The problem here is that data extractors that rely on the Web assume that it is a neutral and democratic place where everyone has a say. For instance, counterarguments to points being made that social media allows people to spread misinformation support the idea of Web neutrality. It is true that people such as Alex Jones were allowed to post online videos arguing that the Sandy Hook mass shooting never

[41] https://www.congress.gov/event/116th-congress/house-event/LC64156/text?s=1&r=3.

[42] https://www.congress.gov/bill/117th-congress/senate-bill/1265.

[43] https://constitution.congress.gov/constitution/amendment-4/.

happened and that the people involved were actors (he was found liable for damages in three defamation lawsuits and sentenced to pay nearly US$1 billion to the families of the victims). But supporters of the "democratic Web" assert that this is the price we need to pay to give everyone the possibility to express their opinion, something not possible before the advent of Internet and the Web. But the Web is, by definition, something that poorly represents the world in which we live. For instance, as Bender et al. (2021) rightly exposed, "size doesn't guarantee diversity," and I would add inclusion. What is most considered by data extractors is webpages in English, and this reflects an overwhelming prevalence of Western culture. Internet contributors are mostly younger users and from the Global North. More specifically, Web scraping often digs into discussion forums such as Reddit, which features contributions mainly from men (67%) aged between 18 and 29 years old (64%) (Bender & Friedman, 2018). Web scrapers also commonly target Wikipedia (especially when one needs relatively polished text to feed generative AI). But females make only between 8 and 15% of contributions on Wikipedia.[44] Not all users can access user-generated content more generally either. Some may even be barred from contributing on a racial basis, as reported by Emily Bender and colleagues[45]:

> [Leslie Key] Jones documents…multiple cases where people on the receiving end of death threats on Twitter have had their accounts suspended while the accounts issuing the death threats persist. She further reports that harassment on Twitter is experienced by "a wide range of overlapping groups including domestic abuse victims, sex workers, trans people, queer people, immigrants, medical patients (by their providers), neurodivergent people, and visibly or vocally disabled people." The net result is that a limited set of subpopulations can continue to easily add data, sharing their thoughts and developing platforms that are inclusive of their worldviews; this systemic pattern in turn worsens diversity and inclusion within Internet-based communication, creating a feedback loop that lessens the impact of data from underrepresented populations.

In addition, in a 2023 article in *The Atlantic*, journalist Alex Reisner reported on a dataset of over 191,000 books that mainly reflect Western

[44] https://rc.library.uta.edu/uta-ir/handle/10106/29572.

[45] https://medium.com/@agua.carbonica/twitter-wants-you-to-know-that-youre-still-sol-if-you-get-A-death-threat-unless-you-re-a5cce316b706.

culture and that were used (without permission) by Meta and Bloomberg (among other companies).[46] Besides the fact that these companies are at the center of several copyright-based lawsuits,[47] the issue is about the lack of control, regulations, and discipline of Web scraping where materials are randomly appropriated by data-hungry companies. The conclusion here is that data extractors are highly biased and can easily lead to DEI issues. Data extractors relying on unsupervised Web scraping simply risk mirroring racial, gender, and disability discrimination that random people post online and are processed by notoriously data-hungry AI systems, which an Electronic Privacy Information Center[48] (EPIC) report in 2023 report on AI potential to generate harms has broadly illustrated.[49] Since many countries (such as America) have insufficient laws and regulations on data scraping, some companies are taking their own action to solve intellectual property issues. For instance, news outlets such as *The New York Times* and CNN have activated systems to prevent ChatGPT's Web crawler from being trained using the content of their webpages.[50] As of early August 2023, *The New York Times* does not allow "the development of any software program, including, but not limited to, training a machine learning or artificial intelligence (AI) system without consent.[51]" While these independent initiatives signal that something can be done at the company level, I make two important observations. The first observation is that these types of "private" actions illustrate the surviving Reganian *laissez faire* approach that the burden of taking care of one's own privacy (or security) is on private individuals or companies because the government doesn't provide such protections. The second observation concerns discrimination that still stems from *laissez faire*—only some individuals can "buy" their privacy (e.g., because they don't use social media since they don't need to find a job via LinkedIn). But, more generally, Web

[46] https://www.theatlantic.com/technology/archive/2023/09/books3-ai-training-meta-copyright-infringement-lawsuit/675411/.

[47] https://fingfx.thomsonreuters.com/gfx/legaldocs/lbpgolxxmpq/META%20AI%20COPYRIGHT%20LAWSUIT%20complaint.pdf.

[48] https://epic.org.

[49] https://epic.org/wp-content/uploads/2023/05/EPIC-Generative-AI-White-Paper-May2023.pdf.

[50] https://www.theguardian.com/technology/2023/aug/25/new-york-times-cnn-and-abc-block-openais-gptbot-web-crawler-from-scraping-content.

[51] https://help.nytimes.com/hc/en-us/articles/115014893428-Terms-of-Service.

scraping (and associated privacy issues) is an example of how large companies (or rich, privileged individuals) can compromise the privacy of only certain individuals.

In conclusion privacy and discriminations affect just some—the weak, the poor, the marginalized, and those who cannot voice their opinion besides the fact that high-tech companies want us to believe that they provide platforms for democracy, free speech, and so on. Online freedom has a price. The currency, besides money, is being male, straight, educated, and White while living in the Global North and having connections and, most importantly, Internet access. Too often we read from the press or legit polling companies that *Americans think this and that*, while forgetting that, to be able to respond to a survey, one needs to have a landline or an email and be enough proficient in English. But, according to the National Center for Education Statistics,[52] 92% of American adults have at least level 1 literacy (people with this level of literacy have difficulty using or understanding print materials). This figure alone means that 26.5 million Americans are not included in what Americans think. The online world follows suit: Web scrapers can ignore the difference between public and available data, and algorithms that are, for the most, not transparent often process such data.

The Future of Data Extractors

The future of data extraction relies on generative AI, which I broadly discuss later in the book. But still, it is worth introducing this innovative technology that will impact some businesses and societies at least as much as Internet and social media did. Generative AI can generate text, images, and other media in response to prompts by anyone with Internet access and a browser or an app. But there's a catch that concerns data extractors: because of the biased nature of Internet, data extractors will feed software such as ChatGPT with data coming from just some sources. In her book *Algorithms of Oppression*, Safiya Noble (2018) envisioned how generative AI can now lead to discrimination triggered by user-generated content. Her book focuses mainly on how Google's algorithms operate when searching the Web and provide supposedly reliable results. These results can be highly biased and discriminatory because the underlying

[52] https://nces.ed.gov.

search engines cannot mitigate the biases of people making queries. For instance, as noted by Latanya Sweeney, the Daniel Paul Professor of the Practice of Government and Technology at Harvard University, Google ranks search results in a way that reflects what the majority of Google users search. In all this, what I found illuminating In Noble's book is her discussion concerning Google's monopoly of information.

The US Federal Trade Commission (FTC) has recently focused on net neutrality, that is, Internet providers' ability (or not) to decide how to prioritize Internet traffic, such as by deciding to slow down access to websites that do not pay them a fee or by giving full speed to websites that do pay a fee. This is certainly an important issue but, as Noble argues, how about content prioritization? Isn't it something concerning net neutrality? Rather than focusing too much on traffic-routing discrimination, wouldn't it be worthwhile to focus more on content prioritization *and* discrimination? The future of data extraction concerns precisely these issues. How will data extractors be able to mitigate current biases affecting Internet searches? How would new search engines powered by generative AI avoid falling into the Google trap? With the advent of generative AI, search engines are being designed in "user-friendly" ways at the point that they will give us answers rather than offering links to websites. Generative AI revolves around the concept that it can generate content (like user-generated content), but, within the realm of search engines, generative AI focus on providing answers to questions that do not involve a list of links but, instead, a verbose answer. This is, in theory, better because we don't need to go over several websites to make up our mind and find the answer to our question. Instead, generative AI will give us an answer. But the issue here is that we have no way to fact check the answer because we cannot vet the sources/links (and we know that generative AI frequently "cheats," something known as AI hallucination).

The future of extraction is a nebulous one in which generative AI systems capture data from people asking questions and share their questions with others. Companies such as Apple, wary of leaks,[53] have restricted employees from using generative AI because their questions feed AI and could appear as someone else's answer. In April 2023, Italy banned ChatGPT after the Italian Data Protection Authority accused ChatGPT of unlawfully collecting users' data and failing to prevent

[53] https://www.wsj.com/articles/apple-restricts-use-of-chatgpt-joining-other-companies-wary-of-leaks-d44d7d34.

underage users from accessing inappropriate material.[54] To wit, privacy concerns associated with generative AI are real, and, as we've already seen, privacy issues concerning data extractors also concern DEI. In fact, massive amounts of data collected by companies can easily lead to discrimination (e.g., the way law enforcement in the US purchases data from brokers to target specific individuals). Also, data collectors do not extract data from all individuals in an equal manner. Some individuals will always have the luxury to escape data collection, which the automotive industry illustrates. If one can afford a full car insurance premium, they won't need to have an onboard device installed in their car. An inclusive society should treat privacy rights of all individuals in the same way and should avoid practices by which some can escape surveillance—or, better still, surveillance should not happen in the first place.

AI also affects facial recognition systems. Besides the overwhelming evidence that facial recognition systems are biased and highly discriminatory (Crawford, 2021a), these technologies are systematically used by law enforcement to make arrests. In 2021, Facebook shot down its facial recognition system that, without user authorization, took advantage of people tagging other people in photos uploaded on the social network (yes, before 2021, Facebook made use of posts to train its AI facial recognition system). A 2021 press release on facebook.com reads as follows:

- *We're shutting down the Face Recognition system on Facebook. People who've opted in will no longer be automatically recognized in photos and videos and we will delete more than a billion people's individual facial recognition templates.*
- *This change will also impact Automatic Alt Text (AAT), which creates image descriptions for blind and visually-impaired people. After this change, AAT descriptions will no longer include the names of people recognized in photos but will function normally otherwise.*
- *We need to weigh the positive use cases for facial recognition against growing societal concerns, especially as regulators have yet to provide clear rules.*[55]

[54] https://www.theverge.com/2023/4/28/23702883/chatgpt-italy-ban-lifted-gpdp-data-protection-age-verification.

[55] https://about.fb.com/news/2021/11/update-on-use-of-face-recognition/.

In Massachusetts, where I live, the state's lawmakers are currently put together a bipartisan state bill that seeks to limit police use of facial recognition systems. This is good news even if the bill's approval hasn't happened yet. In addition, as noted by Tate Ryan-Mosley in a 2023 article in *MIT Technology Review*,[56] regulating how facial recognition systems work might legitimize its use. For instance, if its use is approved only in cases involving violent crimes, the initial outcome will be that law enforcement will not be able to randomly scan people's faces across the state. However, this also means that, for specific crimes, the technology will be seen as appropriate and, thereby, wash out all concerns about its fallacy. Moving from the state to the federal level, a 2022 report issued by the US Government Accountability Office[57] indicates that, while the Federal Bureau of Investigation (FBI) has used facial recognition systems broadly for criminal investigations, only 5% of the workforce has undertaken training to use these systems. This is highly problematic because, given that these systems are far from perfect, not knowing how to use them responsibly might lead to erroneous arrests and prosecutions.

Generative AI will also play a role in the future of facial recognition. OpenAI[58] has already recognized generative AI's ability to look at the world and describe, in a verbose fashion, what it sees. As generative AI can generate images from prompts, it can also "tell" what an image represents. But this aspect of data extractors is problematic because it will be affected by the same shortcomings that currently affect facial recognition systems. The issue will be exacerbated if generative AI is programmed to not just describe what it sees but if it is also asked to make judgments. For instance, a generative AI system used in airports to scan faces of people boarding planes might provide prompts to agents in the Transportation Security Administration (TSA) (i.e., the US agency that oversees security in airports, train stations, etc.) to explicitly target people of color (because, for an AI, it is often problematic to analyze them). In addition, these systems might provide TSA agents with insights on facial emotions (e.g., they could explain to an agent that someone seems nervous and, thus, highlight them as a potential suspect for an in-depth search). But

[56] https://www.technologyreview.com/2023/07/20/1076539/face-recognition-massachusetts-test-police/.

[57] https://www.gao.gov/assets/gao-23-105607.pdf.

[58] https://openai.com.

this would discriminate against emotional individuals given that the literature is rather in agreement in asserting that facial recognition systems cannot capture people's emotions (Crawford, 2021b).

It is, however, worth noting that this generative AI feature could represent an advance in the realm of technologies promoting DEI. For instance, in 2023, an experiment was conducted with visually impaired people who took advantage of generative AI's ability to describe the world with surprising accuracy, which made their life easier. As reported by *The New York Times*, Jonathan Mosen, a blind employment agency chief executive, used visual analysis to determine which dispensers in a hotel room bathroom were shampoo, conditioner, and shower gel. According to his account, this went far beyond the performance of image analysis software he had used in the past.[59] He said: "It told me the milliliter capacity of each bottle. It told me about the tiles in the shower.... It described all of this in a way that a blind person needs to hear it. And with one picture, I had exactly the answers that I needed." This example illustrates the fact that any technology, let alone the most cutting-edge AI developments such as generative AI, can be used for good or for bad. The problem is that, because AI is developed by for-profit companies, the goals remain very much tied to gaining profit and revenue. And, clearly, improving a customer service support with ChatGPT-powered chatbots generates more revenue than developing a similar system for people with disabilities such as the example of Mr. Mosen, which requires more time-to-market effort, approval from the Food and Drug Administration (FDA), negotiation with healthcare providers, and uncertain revenue streams.

An Example of Ambivalent Data Extraction in the Global South

I recently had very interesting conversation with my friend and colleague Silvia Masiero, Associate Professor of Information Systems at the University of Oslo. Silvia is best known for her research on social justice and the role played by technologies, especially in financially suppressed areas and countries, such as the Global South. I asked her to share with me

[59] https://www.nytimes.com/2023/07/18/technology/openai-chatgpt-facial-recognition.html.

some thoughts about data-extraction practices in the Global South, and, in particular, I asked her about lessons learned from her longitudinal study on digital ID in India (see more on her personal blog: https://unfairid.com).

Since 1997, India has attempted to undertake anti-poverty actions by implementing a system revolving around cards given to citizens to allow them to purchase basic needs (mainly food). While everyone in India is entitled to a card (which serves as a universal ID card), only people in need (hundreds of millions) use it to buy food.[60] The system worked relatively well until India started a top-down digitalization process, which included issuing these cards only in digital format. While having a digital ID should create efficiencies and limit issues related to people losing their card in addition to ensuring a match between the digital ID and the card-holder (increasing security), this system poses privacy problems as now everyone's income is recorded in a centralized database. An extract from the website of the public distribution system (PDS) appears below[61]:

> The Public Distribution System (PDS) evolved as a system of management of scarcity through distribution of foodgrains at affordable prices. Over the years, PDS has become an important part of Government's policy for management of food economy in the country. PDS is supplemental in nature and is not intended to make available the entire requirement of any of the commodities distributed under it to a household or a section of the society. PDS is operated under the joint responsibility of the Central and the State/UT Governments. The Central Government, through Food Corporation of India (FCI), has assumed the responsibility for procurement, storage, transportation and bulk allocation of food grains to the State Governments. The operational responsibility including allocation within State, identification of eligible families, issue of Ration Cards and supervision of the functioning of Fair Price Shops (FPSs) etc., rest with the State Governments. Under the PDS, presently the commodities namely wheat, rice, sugar and kerosene are being allocated to the States/UTs for distribution. Some States/UTs also distribute additional items of mass consumption through the PDS outlets such as pulses, edible oils, iodized salt, spices, etc.

[60] Note that this system was created in 1965, but it was only in 199 that the system "targeted" the poor (http://etheses.lse.ac.uk/950/).

[61] https://nfsa.gov.in/portal/pds_page.

Here, once again, we face tradeoffs concerning positive and negative aspects of classification (of the population's poverty status) along with the extraction and digitalization of this information. On paper, India's digital ID is helpful because people below the poverty line can purchase food with financial help from the government in a rather straightforward process. They head to specific stores that are equipped with digital ID readers (most stores, especially in depressed areas of the country) and purchase food. However, what broad consequences would this massive data-collection effort have with respect to individuals' personal information? And what is exactly stored on the digital ID? As Silvia told me, one first important thing to know is that the card is not personal but instead a household card that includes information of anyone living under the same roof (including minors) (Sriraman, 2018). The fact that minors are included is important because it reflects the needs of a family. Information stored in the card include apparently innocuous information such as names, occupations, and income but also more sensitive information such as biometrics (fingerprint scans of ten fingers, retina scans of two eyes, and photographs[62] for each member of the household who is 18 or older). Interestingly, according to Silvia's fieldwork, cardholders are not very concerned with privacy. Several people with whom she spoke either trusted the government or simply didn't bother about privacy issues because "either we give the data away or we don't get food," as Ankita (pseudonym), a study participant in Silvia's fieldwork, indicated. One thing worth noting of this digital system is that the Indian Government, unlike other governments, does not rely on third-party private companies to manage the extensive database containing population data. The technology is extremely centralized and relies on a specific agency (the Unique Identification Authority of India (UIDAI)[63]) that manages all IT operations concerning Aadhaar.[64] This heavily centralized data-extraction system has pros and cons. Pros concern the fact that it is unlikely that the data will be used for commercial use (e.g., by data brokers). In contrast to what happens in the US, where it is mainly private companies that extract data from citizens with various sensor-based extractors

[62] https://anderson-review.ucla.edu/wp-content/uploads/2022/04/Digital-Identity-in-India-Palgrave-Handbook-of-Technological-Finance.pdf.

[63] https://uidai.gov.in/en/about-uidai/unique-identification-authority-of-india.html.

[64] https://uidai.gov.in/en/.

(as I explain above), in India, it is the central government that obtains this data from citizens with less possibility that it will sell the data to third parties. Yet, having all these data in a single database poses obvious security issues along with issues associated with the concentration of power. India has a wide control over people by owning their personal information electronically.

India's digital ID is one such example of a government-led data-extraction initiative that focuses on supporting citizens below poverty line; yet, it also gives India a broad control over people. Similar to a previous example of car insurance companies and people being able to "buy" their privacy by paying a higher premium (and declining to have onboard devices installed on their cars), here people who don't want to be controlled by the government can refuse to become cardholders—especially if they don't need it to buy food. Data extractors and extraction practices have always a bright and dark side, and potential discrimination and privacy issues, along with DEI issues more broadly, can always surface. I believe that a first step to limit these issues is to document how extractions occur and weigh the benefits and potential negative consequences.

Conclusions

Data extractors are extremely powerful technologies not necessarily designed to jeopardize people's privacy or generate discrimination (with exceptions that, to me, relate mainly to facial recognition systems, which so far still need to prove their social utility). For instance, biometric systems might help people lose weight, keep their blood pressure under control, and so on. Web scraping technologies can help social scientists study social phenomena even if they pay poor attention to the difference between public and available data. Even very questionable AI advances such as generative AI can be used for good, such as in the example of people with visual disabilities. But the practices associated with how these technologies are employed by businesses are often red flags. Companies are revenue-driven and use insights from data extractor at their advantage. For instance, car insurance companies ideally want a scenario where they have no incidents they need to deal with. If the people they insure are responsible for an incident, the companies have to pay for it. If people they insure are not responsible for but still involved in an incident, the companies still incur administrative expenses associated with it. If no one has incidents, insurance companies just get the money from their customers (premium). The same happens to HMOs. If a health insurance's patients are all healthy, they won't need medications and they won't receive hospital bills to which HMOs make copayments. So why not rely on data extractors to figure whether a driver is more prone to get into an accident because of their reckless driving and, therefore, raise their premium even before the accident happens (something that resembles predictive policing, often considered a racially based phenomenon)? Or why not rely on them to determine whether an HMO customer has hypertension, in which case the health insurance can decide not to cover that patient anymore since hypertension can lead to costly long-term care associated with stroke and heart attack treatments and lifelong medications? Acting in this raises two problems. First, companies take advantage of data extractors, often without the authorization of the data generators, to make (more) money. Second, these actions generally affect the poor, the marginalized, and, specifically concerning healthcare, Black Americans (for historical reasons explained earlier). It is a pity that data extraction is exploited in such a way because data extractors can be used for good. Especially in healthcare, sharing data collected from large populations can lead to the discovery of disease treatments and even to the discovery of preventive care. The issue here is that, when predictions generated by data extraction contribute to large high-tech businesses' bottom line, these businesses try to circumvent current laws and regulations to acquire and analyze these data. When instead these predictions lead to long-term, uncertain benefits (such as improving the health of populations), companies refrain from identifying exploitable loopholes. For instance, companies selling wearable devices do not shy away from capturing and selling biometric data for marketing purposes (i.e., offering a smartwatch owner ads for blood pressure medications should the owner showcase high blood pressure readings). Yet, the same companies do not question the rigid policies about protecting health data when sharing such data could lead to research aimed at curing or preventing diseases. When it is governments that extract data, such as in the case of India, slightly different considerations apply. On the one hand, governments' owning so much data on their citizens in digital format reflect power, control, and potentially oppression. On the other hand, knowing more about their citizens' income, family conditions, and so on might help governments target specific aid to whoever needs it the most.

References

Barnard, C. I. (1938). *The Functions of the Executive*. Harvard University Press.

Bender, E. M., & Friedman, B. (2018). Data Statements for Natural Language Processing: Toward Mitigating System Bias and Enabling Better Science. *Transactions of the Association for Computational Linguistics, 6*, 587–604.

Bender, E. M., Gebru, T., McMillan-Major, A., & Shmitchell, S. (2021). On the Dangers of Stochastic Parrots: Can Language Models Be Too Big? *Proceedings of the 2021 ACM Conference on Fairness, Accountability, and Transparency* (pp. 610–623).

Beyer, M. A., & Laney, D. (2012). *The Importance of Big Data: A Definition* (Gartner Report, pp. 1–9).

Bridges, K. (2011). *Reproducing Race: An Ethnography of Pregnancy as a Site of Racialization*. University of California Press.

Crawford, K. (2021a). *The Atlas of AI: Power, Politics, and the Planetary Costs of Artificial Intelligence*. Yale University Press.

Crawford, K. (2021b). Time to Regulate AI That Interprets Human Emotions. *Nature, 592*(8), 167.

Curchod, C., Patriotta, G., Cohen, L., & Neysen, N. (2019). Working for an Algorithm: Power Asymmetries and Agency in Online Work Settings. *Administrative Science Quarterly, 63*(5), 644–676.

Erlich, Y., Shor, T., Pe'er, I., & Carmi, S. (2018). Identity Inference of Genomic Data Using Long-Range Familial Searches. *Science, 362*(6415), 690–694.

Faraj, S., Pachidi, S., & Sayegh, K. (2018). Working and Organizing in the Age of the Learning Algorithm. *Information and Organization, 28*(1), 62–70.

Gal, U., Jensen, T. B., & Stein, M.-K. (2020). Breaking the Vicious Cycle of Algorithmic Management: A Virtue Ethics Approach to People Analytics. *Information and Organization, 30*(2), 1–15.

Huselid, M. A. (2018). The Science and Practice of Workforce Analytics: Introduction to the HRM Special Issue. *Human Resource Management, 57*(3), 679–684.

Jackson, P. (1986). *Introduction to Expert Systems*. osti.gov. https://www.osti.gov/biblio/5675197

Jhaver, S., Karpfen, Y., & Antin, J. (2018). Algorithmic Anxiety and Coping Strategies of Airbnb Hosts. *Proceedings of the 2018 CHI Conference on Human Factors in Computing Systems*. ACM, paper 421.

Kellogg, K., Valentine, M., & Christin, A. (2020). Algorithms at Work: The New Contested Terrain of Control. *Academy of Management Annals, 14*(1), 366–410.

Lee, M. K., et al. (2015). *Working with machines: The impact of algorithmic and data-driven management on human workers. Proceedings of the 33rd Annual ACM Conference on Human Factors in Computing Systems, Seoul,* South Korea, ACM.

Lepore, J. (2020). *If Then: How the Simulmatics Corporation Invented the Future*. Liveright Publishing.
Lewis, D. (2023). China's Souped-up Data Privacy Laws Deter Researchers. *Nature*. Published on March 25, 2021. https://www.nature.com/articles/d41586-023-01638-1. Accessed on July 15, 2023.
Loebbecke, C., & Picot, A. (2015). Reflections on Societal and Business Model Transformation Arising from Digitization and Big Data Analytics: A Research Agenda. *The Journal of Strategic Information Systems, 24*(3), 149–157.
Lyytinen, K., & Yoo, Y. (2002). Ubiquitous Computing. *Communications of the ACM, 45*(12), 63–96.
Marabelli, M., Hansen, S., Newell, S., & Frigerio, C. (2017). The Light and Dark Side of the Black Box: Sensor-Based Technology in the Automotive Industry. *Communication of the AIS, 40*(16), 351–374.
Marabelli, M., & Markus, M. L. (2017). Researching Big Data Research: Ethical Implications for Is Scholars. *Americas Conference of Information Systems (AMCIS)*, Boston, MA.
Marabelli, M., Newell, S., & Handunge, V. (2021). The Lifecycle of Algorithmic Decision-Making Systems: Organizational Choices and Ethical Challenges. *Journal of Strategic Information Systems, 30*, 1–15.
Marabelli, M., Zaza, S., Masiero, S., Li, J., & Chudoba, K. (2023). Diversity, Equity, and Inclusion in the AIS: Challenges and Opportunities of Remote Conferences. *Information Systems Journal, 33*(6), 1370–1395.
McAfee, A., & Brynjolfsson, E. (2012). Big Data: The Management Revolution. *Harvard Business Review, 90*(10), 61–67.
Mintzberg, H. (1980). Structure in 5's: A Synthesis of the Research on Organization Design. *Management Science, 26*(3), 322–341.
Möhlmann, M. (2021). Algorithmic Nudges Don't Have to Be Unethical. *Harvard Business Review*. https://hbr.org/2021/04/algorithmic-nudges-dont-have-to-be-unethical
Möhlmann, M., De Lima, A., Salge, C., & Marabelli, M. (2023). Algorithm Sensemaking: How Platform Workers Make Sense of Algorithmic Management. *Journal of the Association for Information Systems, 24*(1), 35–64.
Möhlmann, M., & Henfridsson, O. (2019, August 30). What People Hate About Being Managed by Algorithms, According to a Study of Uber Drivers. *Harvard Business Review*.
Möhlmann, M., Zalmanson, L., Henfridsson, O., & Gregory, R. W. (2021). Algorithmic Management of Work on Online Labor Platforms: When Matching Meets Control. *MIS Quarterly, 45*(4), 1999–2022.
Newell, S., & Marabelli, M. (2015). Strategic Opportunities (and Challenges) of Algorithmic Decision-Making: A Call for Action on the Long-Term Societal Effects of 'Datification.' *The Journal of Strategic Information Systems, 24*(1), 3–14.

Noble, S. U. (2018). *Algorithms of Oppression*. New York University Press.

Nunan, D., & Di Domenico, M. (2022). Value Creation in an Algorithmic World: Towards an Ethics of Dynamic Pricing. *Journal of Business Research, 150*, 451–460.

Sendak, M. P., Ratliff, W., Sarro, D., Alderton, E., Futoma, J., Gao, M., Nichols, M., Revoir, M., Yashar, F., & Miller, C. (2020). Real-World Integration of a Sepsis Deep Learning Technology into Routine Clinical Care: Implementation Study. *JMIR Medical Informatics, 8*(7), 1–16.

Seto, E., Challa, P., & Ware, P. (2021). Adoption of Covid-19 Contact Tracing Apps: A Balance Between Privacy and Effectiveness. *Journal of Medical Internet Research, 23*(3), e25726.

Sommers, S. R., & Marotta, S. A. (2014). Racial Disparities in Legal Outcomes: On Policing, Charging Decisions, and Criminal Trial Proceedings. *Policy Insights from the Behavioral and Brain Sciences, 1*(1), 103–111.

Sriraman, T. (2018). *In Pursuit of Proof: A History of Identification Documents in India*. Oxford University Press.

Taylor, F. W. (1911). *The Principles of Scientific Management*. Harper & Brothers Publishers.

Tursunbayeva, A., Di Lauro, S., & Pagliari, C. (2018). People Analytics—A Scoping Review of Conceptual Boundaries and Value Propositions. *International Journal of Information Management, 43*, 224–247.

Wolfsfeld, G., Segev, E., & Sheafer, T. (2013). Social Media and the Arab Spring: Politics Comes First. *The International Journal of Press/Politics, 18*(2), 115–137.

Zuboff, S. (2019). *The Age of Surveillance Capitalism: The Fight for a Human Future at the New Frontier of Power*. PublicAffairs.

CHAPTER 3

Training AI, Computation, and the Environment

In this chapter, I collapse the insights put forward in the previous two chapters about classification and data extractors by exploring *how* data extracted from data generators (so, human beings) via sensors and Web scraping is being classified. This happens both manually and automatically with techniques aimed at teaching or, better, training AI on how to use data and classifications through computations that lead to outcomes that support decision-making. Manual training concerns data labeling where people manually teach AI (e.g., what is embedded in a photo, such as an elephant, a chair, etc.). Automated training is about AI learning while "in production," something typical of ML, that improves algorithms as they are used in practice. Training AI is a key task that is embedded in the whole AI lifecycle (design, implementation, and use in practice) as I show in a recent paper (Marabelli et al., 2021a). However, training comes at a price. Manual training is often done by people not qualified for the job (not their fault, but it relies on cheap work and exploitation). Automated training is risky because it is often done using notoriously biased historical data—think about what would happen to an AI hiring system that one trained to select potential executives at Fox News based on historical data of successful people (e.g., their gender, age, race, etc.). Also, automated training requires complex algorithmic work (computations). Automated training, when an AI system is "in production," relies

M. Marabelli, *AI, Ethics, and Discrimination in Business*, Palgrave Studies in Equity, Diversity, Inclusion, and Indigenization in Business,
https://doi.org/10.1007/978-3-031-53919-0_3

on live data from whoever uses the system. For instance, generative AI, as discussed earlier, learns from people's prompts. While I mention it due to privacy issues in the previous chapter, here I focus on biases that can be injected into AI systems by users—something not new to algorithmic scholarship as it has happened for a long time with Google's search engine (Noble, 2018). Also, training AI models (especially ML models) requires an unbearable amount of machine resources, which turns to be detrimental to the environment because machine work requires cooling and electrical power, both scarce resources, and their exploitation can harm the environment. Given the undeniable climate changes we witness nowadays, it is of paramount importance to focus on the environment when discussing training.

The DEI implications of all this are remarkable, and worrying. Manual AI training is based on offshoring exploitation, which is highly unethical. What is more, manual training of AI (e.g., based on labeling photos) is highly biased because the people involved are not suited for the job. In this chapter, I include content moderation as a form of manual and automated training. To this end, I explain how miserable content moderators are and the associated emotional stress they experience. Needless to say, content moderators are exploited people and, similar to other manual moderation tasks, often outsourced.

Automated training requires algorithm design that prevents AI from being used with biased data, which is challenging given the lack of diversity in AI development teams at major high-tech companies. When automated training doesn't use live data, it uses Internet data and, thus, suffers from diversity issues that are inherent to the Internet (Bender et al., 2021). One important risk that requires examination is associated with how Internet content is being populated nowadays given the advent of generative AI. The more AI-based (and biased) content populates the Internet, the more we risk training AI systems with their own data—something that I call *AI Cannibalism Disease*. Finally, the environmental consequences of automatically training AI models are unequal across populations with poor countries generally hit harder by climate change (and with less resources to face these changes) even though they minimally contribute to global warming. America unfortunately exemplifies negative effects that can result from automated training of AI models without sufficient care. But things are changing, and EU-led initiatives are fueling US initiatives to acknowledge AI's energy consumption.

Manual Training

AI systems require thousands of human work hours to behave in a way their creators want them to, and even then, they do it in an unreliable manner. But what is exactly manual training and who are the actors performing such activities? While seldom discussed, the role of humans in system training is at least as important as the human in the loop or any other human role when AI is being used. While AI systems "in production" can be partially self-trained (especially in the case of neural networks), there is a whole aspect of training that is heavily led by humans—and this happens at any stage of the AI lifecycle (development/ design, deployment, and use in practice). This aspect of AI training is critical because individuals involved often lack the competencies to train AI, are exploited and surveilled, and, in some cases, act unprofessionally by sharing training data about people without their consent. Offering a few examples will help clarify these points.

As I noted earlier, AI systems can be used to recognize images of individuals (i.e., facial recognition systems). They are not reliable, and their performance decreases when attempting to recognize people with darker skin and women (Smith & Miller, 2021). For instance, they have led to the wrongful detention of individuals.[1] But the technical shortcomings of facial recognition systems often result from design issues and flaws that have to do with how the systems are trained. In my opinion, a key aspect of training facial recognition systems that deserves attention (and that is, sadly, largely overlooked in ML classes) concerns the so-called labeling process in which individuals "tag" characteristics of images to train these systems on how to recognize similar images or even live streams (i.e., identifying a person from a set of properly labeled photos). First, labeling is developed more intensively on photos of white people (particularly white males). Second, the labeling process often relies on people who did not receive training to do so.

The first issue is rooted in the white male-dominated culture of technology developers of the past half century—a good historical account is provided by Safiya Noble (2018). AI development is in the hands of large, revenue-driven tech companies. And White males are typically not only the owners and relevant stakeholders of these companies but also the most

[1] https://www.npr.org/2020/06/24/882683463/the-computer-got-it-wrong-how-facial-recognition-led-to-A-false-arrest-in-michig.

profitable customers. Consequentially, AI systems have been systematically fed with "related" data. This way of using training data in a selective (and racist) way has been rather common in healthcare clinical trials in the US for almost a century (e.g., only white people were invited to participate in studies) such that drugs, but in particular medical devices were designed to work better with white people (Obermeyer et al., 2019). It has also culminated in a number of explicitly racist AI applications that were deployed during the COVID-19 pandemic that made use of skewed data and worked only for certain races (Whites) (Marabelli et al., 2021b). Problems concerning skewed training data are historical and, in America, rest on the fact that healthcare systems were always designed for White people. The private nature of healthcare had the consequences that mostly White people could access it—especially before the Affordable Care Act[2] (also known as Obamacare, enacted in March 2011)—which broadened coverage and focused on financially weak people. This gave access to a more diverse population; in America, White people, as a historically privileged population, make more money and are (were), therefore, able to access (private) care. Figure 3.1, from the US Department of Labor (DOL), shows important race and gender differences between Whites and Blacks in America. White men's median annual earning has been above US$60,000 since 2016, while White women have had a similar trend to Black men—roughly above US$50,000 in the same period. Black women's annual earning has been substantially lower, even if it is worth noting that it jumped from almost $38,000 to $46,500 from 2010 to 2021, most likely because of the Obama administration (even if solid sources supporting this cause-effect relationships are absent).

This data is a good example of the importance of classification. This book considers classification processes to showcase how dangerous is to classify people (especially when we do it based on skin color), but, as consequence of perpetuating discriminations, having data classified in this way can help one realize the magnitude of social injustice. Bottom line, in a male-dominated society, especially in business, it is not surprising that AI systems are designed by and trained for the people in power (Whites). Unfortunately, exposing these issues on government websites such as the DOL website doesn't help the cause. High-tech companies pursuing revenues serve the rich because they are more profitable (i.e., rich people

[2] https://www.hhs.gov/healthcare/about-the-aca/index.html.

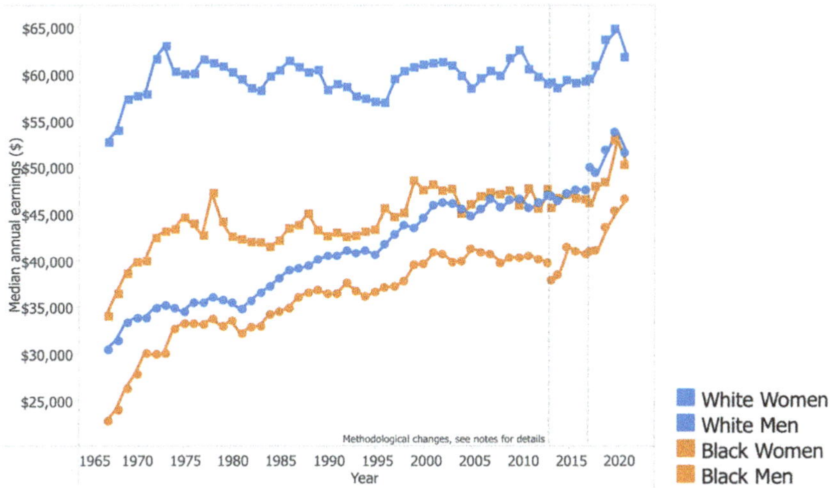

Fig. 3.1 Medial annual earning by gender and race (*Source* DOL[3])

pay medical bills because they have private insurance, because they are employed, and their employers take care of their coverage). A spinoff of this trend is the composition of AI developers and AI researchers more generally. A 2021 report from Stanford University[4] provides rich details on who designs AI systems and the trends in educating data scientists. Specifically, it showcases a lack of women and Black people enrolled in related PhD programs. Some (wrongly) suggest that the lack of gender and race diversity in AI companies is due to the so-called pipeline effect by which, for instance, few women graduating in science, technology, engineering, and mathematic (STEM) programs do not allow companies to hire a fair share of women. If nine out of ten candidates are men, and there are four openings, it is impossible to hire two women and two men.

But this is an extreme case and, generally, women applying for positions as data scientists are "enough" to give recruiters a chance to pursue gender diversity. They should be hired (of course, if qualified) not just

[3] https://www.dol.gov/agencies/wb/data/earnings/median-annual-sex-race-hispanic-ethnicity.

[4] https://aiindex.stanford.edu/wp-content/uploads/2021/03/2021-AI-Index-Report-_Chapter-6.pdf.

because of DEI principles—something that would be anyway a sufficient reason to do so. High-tech companies should hire more female data scientists because diverse teams are proven to produce better outcomes in terms of technical quality and performance and because diverse teams lead to fairer systems (in this case, with respect to potential gender biases) because different actors building a system are sensitive to different issues (especially ones that concern themselves). For instance, in the 70s, car seatbelts were designed without thinking of how they could protect pregnant women perhaps in part because no pregnant women were involved with the design teams (Perez, 2019). This principle, by which we need to recognize diversity (even just because of how people look like), is, however, prone to abuse. Yes, what people look like matters.

For instance, the Tennessee's Student Teacher Achievement Ratio (STAR) Project,[5] a longitudinal study that has occurred in 1985–1990, showcased that Black children learnt better from Black teachers. Another example involves After Dr. Derrick Bell, then a Harvard Law School professor who took an unpaid leave of absence in 1990 to protest against Harvard denying tenure to two Black professors committed to CRT (critical race theory) (Harvard claims they denied tenure to the two professors because they were underperforming). When several Black students protested because Dr. Derrick's summer seminar on CRT was canceled, Harvard offered to hire someone else to teach that seminar. The choice went to White professor Jack Greenberg, then director of the NAACP (National Association for the Advancement of Colored People), a civil right organization. The students protested because they wanted the course to be taught by a Black professor, believing that only someone who experienced "first hand" social injustices against Black communities could effectively led the seminar. How people look like matters, by far, but the downside of hiring strategies relying on affirmative actions is that minorities become the target of passive aggressive bullying practices, where these minorities feel that (White) people around them think they got their job because of their race, and not because of their skills and abilities.

The second issue stems from the fact that data training does not happen automatically; for instance, an ML system needs thousands of photos of an elephant to recognize one at the zoo (note that a toddler

[5] http://www.classsizematters.org/wp-content/uploads/2016/09/STAR-Technical-Report-Part-I.pdf.

Fig. 3.2 Biases in labeling[6]

needs just a couple to perform the same task). Within the realm of image-recognition systems, AI needs to be trained with labeling processes that involve humans who often do not receive training to do so. For instance, people who work on Amazon Mechanical Turk (MTurk) often label photos of countless individuals for less than minimum wage and without having gone through training. Figure 3.2 illustrates an example of inexperienced data labelers asked to place tags on photos of two congressmen. Schwemmer and his team (Schwemmer et al., 2020) conducted a study with MTurk workers asked to perform a typical task required to set up ML systems: labeling various photos. They gave the MTurk workers various photos of US congressmen and found astonishing outputs. They saw males as powerful business leaders and projected the aspect of a man with a suit into their ability to discharge power. Females instead were looked at for their physical characteristics without a strong link between such characteristics and their role in society. Thus, they labeled men with terms such as "business," "speaker," "official" (all activities associated with the photo/person), while they labeled women with terms such as "smile,"

[6] Courtesy of Carsten Schwemmer.

"chin," "black hair," and so on (personal characteristics rather than terms associated with abilities as with the men).

This is very concerning because it is in this way that facial recognition systems are often trained. The consequences and the problematic outcomes form a DEI perspective are obvious but not yet addressed—or at least they are addressed once exposed by news outlets or research papers. For instance, in 2020, the *Wall Street Journal* published an article bout Jacky Alciné, a Web developer, who found that Google's photo app mistakenly tagged two black people as gorillas.[7] Google's apologetic response was followed by the company banning the search term "gorilla" from the app—the patch they were able to apply to the issue at hand. In this case, clearly, the AI system running Google's photo app was trained with an insufficient number of photos of Black individuals, something that is consistent with the overwhelming prevalence of photos of White individuals online. This example differs from the labeling issue above. But both convey the same message: that facial recognition systems are not trained effectively either because labelers are underqualified or because these systems are not fed with enough diverse data. But the problem here goes beyond technical shortcomings. While it is worrisome that large high-tech companies such as Google take such a shallow approach when building AI systems, knowing that they do means it is unacceptable that these systems are unleashed in society to, for example, recognize faces in airports, screen job candidates, and support law enforcement personnel who, based on the outputs of facial recognition systems, make arrests. The lack of inclusiveness of non-White people in training data sadly mirrors our society because we always need to remember that technologies reflect the people who build them.

This is also tied to high-tech companies' focus on efficiency and efforts to bring innovations to the market quickly. It is not a secret that companies such as Meta, Google, and Microsoft (to mention just three of the biggest AI players) believe that vetting innovations such as new AI models and systems for too long slows down progress—a typical capitalistic and consequentialist approach where these companies prefer to roll out innovations to the general public *before* they are fully tested. The actual test occurs when millions of people start using them. This makes sense from a business standpoint because, in this way, user tests (very expensive and

[7] https://www.wsj.com/articles/BL-DGB-42522.

time consuming) are minimized, and a company can make sure to be on the market with the most cutting-edge technologies before others do the same. But this behavior comes at the expense of the end users. In the late 1990s and early 2000s, Microsoft, used to release poorly tested software updates (on Windows 95/98/2000 operating systems) and then, a couple of times a year, release "service packs" or general operating system updates. In this way, it collapsed the patches built upon user feedback related to the quickly released updates. While users probably shouldn't have accepted it, at that time, Microsoft was the undisputed leader (and monopolist) of software for home computers, and it caused no specific harms. The software wasn't flawed because it discriminated against some individuals, nor did it have privacy issues or treat people in unequal ways, and so on. Excel with a bug was a spreadsheet that would crash when the user tried to generate a macro involving multiple cells.

Today, with AI deciding the present (e.g., self-driving cars) and the future (e.g., hiring software) of our everyday life, unleashing poorly tested technologies can have dramatic effects. But this is reality, the quintessence of capitalism, which prevents (or tries to) government interventions and leads to the current laissez faire approach to technology. In this way, high-tech companies in power lead the development and, most importantly, the deployment of poorly trained systems to the general public in the name of private innovation. This laissez faire trend, pursued in recent times by the Reagan administration and, after being relatively dormant for almost four decades,[8] brutally resurrected by the Trump administration, is now affecting the extent to which institutions have and will have a say in technology development. An excellent storytelling of the role of government in America in regulating business operations is *Invisible Hands*, a book by the Gardiner-Kenneth T. Jackson Professor of History at Columbia University Kim Phillips-Fein (2009). The book does not discuss contemporary technologies (it actually doesn't discuss technologies at all—at least as we intend them (i.e., digital)). It does, however, precisely recount the US Government's role in regulating (or not) private businesses since the creation of the New Deal in the 1920s and helps explain why we now face these uncomfortable statements of high-tech leaders who think that deploying potentially biased AI is good for progress and why the government doesn't act to protect people who can be harmed by these

[8] Some argue that even the Clinton and Obama administrations did very little to regulate industries—or, at least, their efforts to do so were mainly pointless.

technologies. I discuss specifically this topic in Chapter 5. But, maybe not surprisingly, poorly tested technologies do not affect just end users. They also affect people who are underpaid and work long hours to contribute to speeding up AI's time to market. These workers, the labelers, are often people from countries in the Global South.

The case of Venezuela is illustrative of how the AI ecosystem is detrimental to DEI, specifically with respect to labelers. The country's economic crisis, since 2018, has led many well-educated, Internet-connected citizens to seek work on crowdsourcing platforms, including data labeling, to make a living.[9] The crisis has been one of the worst economic disasters in the country in nearly 50 years and has caused widespread food shortages, hyperinflation, and a collapse of the health-care and education systems.[10] Many people have turned to online work to support themselves in the face of these challenges. Consequently, this crisis has benefited certain companies by providing access to (or exploit) cheap labor. However, for Venezuelans, the growth of this industry in their country has had both positive and negative effects. On the positive side, it has provided a source of income for people who may not have had other options. On the negative side, it has exposed them to exploitation by corporations that have reduced pay, suspended accounts, or terminated programs to offer low-cost services to tech companies. The situation has created a significant power imbalance with platforms holding control over how things are conducted and the rules of the game, which has created a new model of labor exploitation in the country.

During the COVID-19 pandemic, some companies reduced wages and increased working hours for gig economy workers due to tight budgets and excess labor, which has driven down the average cost of data labeling.[11] Specifically, this situation reduced gig workers' power since it restricted their ability to seek better paying jobs and gave hiring companies the power to lower their wages. As a result, gig workers have seen a decline in income that no longer allows them to support themselves or their families. This trend has been particularly noticeable since, to maintain competitive pricing, these companies often hire workers

[9] https://techcrunch.com/2019/03/29/venezuela-is-losing-A-generation-of-tech-talent-to-its-humanitarian-crisis/.

[10] https://www.cnbc.com/2019/08/02/venezuela-inflation-at-10-million-percent-its-time-for-shock-therapy.html.

[11] https://time.com/5836868/gig-economy-coronavirus/.

from disadvantaged and marginalized groups, such as low-income youth, refugees, people with disabilities, and even minors, as per a recent Wired Magazine investigation explains.[12] That exploitation highlights inequities on the income level and on the access to fair-paying work for these groups.

A growing concern is that using cheap labor from vulnerable countries, similar to the exploitation of colonial times, is spreading as some platforms seek even cheaper labor in other countries. The model, which originated in Venezuela, has set low expectations for payment in the AI industry and has established a strategy for meeting the low prices demanded by clients. This phenomenon is possible due to a combination of poverty and good infrastructure and is spreading across the world. For instance, to build ChatGPT, starting in November 2021, OpenAI outsourced its labeling operations to Kenyan laborers earning less than US$2 per hour. In this case, the goal was not to simply train AI to recognize photos. While building ChatGPT (fed mostly with data available online), OpenAI realized that some of the software's outputs were "problematic." As reported by *Time* in January 2023[13]:

> The AI had been trained on hundreds of billions of words scraped from the internet—a vast repository of human language. That huge training dataset was the reason for GPT-3's impressive linguistic capabilities, but was also perhaps its biggest curse. Since parts of the internet are replete with toxicity and bias, there was no easy way of purging those sections of the training data. Even a team of hundreds of humans would have taken decades to trawl through the enormous dataset manually. It was only by building an additional AI-powered safety mechanism that OpenAI would be able to rein in that harm, producing a chatbot suitable for everyday use.

To generate an AI-powered safety mechanism, OpenAI replicated what social media companies such as Facebook had done in the past to prevent hate speech and toxic language using automated tools. The technique used by social media companies involved feeding AI aimed at detecting hate speech with labeled examples of toxic language and hate speech. Then, the tool would learn how to detect similar online

[12] https://www.wired.com/story/artificial-intelligence-data-labeling-children.

[13] https://time.com/6247678/openai-chatgpt-kenya-workers/?mc_cid=7b2f7f31b4&mc_eid=687b492165.

contributions automatically. In turn, OpenAI built a dataset with myriad examples of child sexual abuse, bestiality, murder, suicide, torture, self-harm, and incest and outsourced the labeling process to Kenyan workers via a San Francisco-based company (Sama) that employed workers in Kenya, Uganda, and India. To wit, to build a tool that companies and individuals in the Global North can use to boost customer care with chatbots, support law practices generate routine documents, and help college students learn faster by summarizing papers and literatures (or cheat on assignments), just to mention a few examples, AI companies rely on exploiting the weak, the poor, and the marginalized. But these examples of outsourcing labeling processes go beyond incompetence of labelers and financial exploitation (as it was the case of MTurk and facial recognition systems). Here, people (from the Global South) can become emotionally drained due to their exposure to obscenities. But these people are without an alternative: their financial struggles force them to accept these underpaid, temporary job as the only alternative to unemployment.

But OpenAI is just one example of these offshoring, almost invisible practices leading to great outcomes in the Global North at the expense of financially and emotionally targeted people in the Global South—a sign of the inequal ways innovations propagate across the world. In June 2023, a court in Kenya issued a landmark ruling against Meta for exploiting Kenyan workers that it asked to train algorithms for two of its two key platforms: Facebook and Instagram. Similar to OpenAI, Meta required workers to go over obscene content here with the aim to moderate it directly rather than training systems to do so (in this case, via the Luxembourg-based company Majore).[14] The substance, however, is the same. Global South workers are asked to do things no one would unless they could already make ends meet. According to an article in *Wired,* Moderators such as Mr. Daniel Motaung (who was fired after trying to form a union) reported that their job was about "watching hours of violent, graphic, or otherwise traumatizing content daily, left him with post-traumatic stress disorder."[15] I don't want to leave the reader with a sense of dark pessimism concerning data companies and AI workers (even if facts are facts). There are examples of companies, even in

[14] https://www.reuters.com/technology/kenyan-court-orders-mediation-meta-labour-dispute-2023-08-23/.

[15] https://www.wired.com/story/tiktok-leaked-documents/.

the Global South, who now reflect ethical standards. For instance, Billy Perigo recently reported on the Time of Kayra,[16] a non-profit company in Bengaluru, the capital of India's southern Karnataka state, that defines itself as "the world's first ethical data company."[17] The company's website states: "We capture, label and annotate data for our clients with high accuracy, and in doing so, we provide critical, supplementary income to people in rural communities." The company offers much higher wages than peer companies and funnels revenues, once fixed and variable costs are covered, into the rural poor India. This example illustrates the fact that it is possible to offshore AI operations concerning data labeling, annotation, and so on without exploiting weak economies (actually, the Kayra case seems to do the opposite by bringing money to needy populations (rural communities)). However, clearly more needs to be done to spread this business model and persuade large high-tech companies to consider forms of offshoring beyond mere exploitation.

In summary, the worker-exploitation examples in vulnerable countries uncover DEI aspects of training data from a totally unique perspective. First, training data is built in a way that illustrates the power of dominant companies, countries, and cultures at the expense of fair and equitable remuneration. Second, because many workers desperately need to make ends meet, they might conduct the "teaching AI" process in a rush; thus, the issue concerns sloppy jobs by competent people rather than a seemly accurate job by untrained MTurk worker. The two key training activities of labeling and moderating online content to give AI systems inputs on how to do so on their own illustrate the fact that, according to Srravya Chandhiramowuli, a PhD researcher studying labor practices in data work at the University of Edinburg "This notion that you can build these large-scale systems without human intervention is an absolute fallacy."[18] It is important to highlight the role of human beings in building AI systems because the AI economy in on its way to mirror longstanding practices of exploitation relying on offshoring manual work to what we used to call "developing countries" or "emerging economies."

[16] https://www.karya.in/about.html.

[17] https://time.com/6297403/india-ai-karya-startup/.

[18] https://www.technologyreview.com/2023/06/13/1074560/we-are-all-ais-free-data-workers/.

To deepen my understanding on the exploitation of labor in the Global South concerning data labeling and, more generally, on the AI supply chain, I asked Srravya to tell me more about her research, which focuses on the contribution and the role of data labeling and the role that data labelers play in the development of AI systems. Specifically, Srravya is interested in what is the role of the human and what is the contribution of the human worker in making these AI systems possible. A first important take from my chat with Srravya is that labeling is not a one-off process. Companies develop models using labelers, and then need to continuously update these modes with new "real-world" data, which requires labelers. As she noted:

> They're constantly developing their models, making changes to their models, enhancing the features, improving its efficiency. And for each of those things to happen, there is a continued reliance on newer data sets to be provided through data labeling. So they always need more and more annotated and labeled data sets to be able to make these new enhancements, improvements, and changes in efficiency of these AI systems.

This is interesting because it gives a view of the AI supply chain as a complex system with labelers representing an integral part of this system even when a model is deployed and used by end users. Along with labelers employed to enhance the system, a second area where labelers are used in the long-term concerns model verifications; that is, the assessment of the efficiency and effectiveness of specific AI systems. Labelers make sure that the outputs of an AI system are consistent with the expectations of the designers. Srravya told me:

> When a model runs and is used to say, for instance, in a facial recognition system, is it doing the work of identifying faces in a video feed or in a data set of images? They verify the output so that the accuracy of the system is high, and even this verification that they do, yes, this is correct, no, this is not correct, that again gets fed into the system as a further training data.

An example from her own fieldwork concerns automated in-store billing:

> In one of the projects that I've observed, there's a case where it's an automated billing at a store. You walk into a store, you can pick up something

> and exit and you get automatically billed. So in that an experience, how does a customer get billed? It's because there's an AI system in place, which is looking at your camera feed in the store and based on sensors and a lot of other things, determine what products you pick. Now in this project, there's actually 100-odd people sitting in India across three different shifts. They work from 7:00 AM to 4:00 p. M, 1:00 p. M. To 10:00 p. M. And then 10:00 p. M. To 7:00 a. M. In three different shifts, so that any time the AI fails and a customer is built wrongly, they would then step in and say that, no, this person's been billed wrongly. They've picked up only two Cola bottles, they've been billed for four, and they manually correct the billing for the customer.

What is interesting from this example is that AI labelers provide real-time feedback that, while relevant with respect to the reliability of the system itself, becomes extremely stressful for workers. My concerns about exploitation of labelers were partially mitigated by Srravya who told me that many labelers she encountered in India take on these jobs with the hope of one day becoming coders and then data scientists who will design, rather than manually assess, these systems. Though she still admitted that the data labeling ecosystem is highly problematic and a source of exploitation. Data labelers are put under a lot of pressure due to the fact that their tasks are heavily monitored. AI companies need to deliver models labelers need to perform specific tasks in a timely manner with an emphasis on productivity. How many tasks can they complete every day? How many tasks can they complete every hour? There are elaborate spreadsheets maintained by team leads who track every person on their team and how much work they've completed every hour of the work day. Srravya told me:

> The productivity demands, like I mentioned, they're given hourly targets, and because they're given hourly targets, it's working pretty much non-stop for at least eight to nine hours a day, let's say maybe 45 minutes for lunch in between, a 15-minute break here and there in the morning and afternoon, but really that's it.

With all that said, end users don't seem very sensitive to how AI companies harvest data to train their systems (or they just don't know). Interestingly, while revelations of American companies exploiting labor in the Global South have almost often led to customers shifting purchase preferences to more ethical companies (Fig. 3.3 gives an example of a sweatshop in India), exploitation of workers for AI seems to generate no reactions among users. For instance, in the 1990, Nike was accused to have moved its offshoring operations from Korea and Taiwan to other countries due to the increased cost of labor in these two countries. Having moved operations to India, Pakistan, and Indonesia (among other countries), Nike subcontractors allegedly provided extremely poor work conditions. In 1996, *Life* reported child labor in some of Nike's offshored sites,[19] which created a worldwide scandal. Nike subsequently took actions including more oversee of offshore work conditions while releasing a social responsibility report in 2005.[20] The Nike case is just an example of how people are sensitive to whether companies are "ethical." According to a 2018 study on American consumers conducted by Accenture,[21] the majority of respondents said that they prefer to spend their money with companies that act ethically. Yet, the impact of the revelations of how AI models are manually trained is not even the subject of debate (besides niche outlets that I've cited in this chapter). Some explanations follow.

First, consumers seem more addicted to online resources than shoes. And they forget and forgive online companies very easily. For instance, Facebook, in the last decade, has demonstrated that it lacks care for its users' privacy (e.g., the 2018 Cambridge Analytica scandal that I mention in Chapter 2). In 2011, the company conducted a social experiment (known as emotion contagion experiment or Facebook mood experiment) in which it manipulated the news feeds of over 650,000 users to discover whether people who post differently based on whether they saw positive or negative feeds. The experiment revealed that people exposed to mostly negative feeds (people reporting the death of a relative or posting about a war) were more likely to post themselves negative feeds. The opposite happened to people exposed to positive feeds. Facebook conducted this experiment in the wake of debates at the time on whether social media could lead to depression. They wanted to figure out the extent to which their platform could impact people's psychological mood. In the same period, communication researchers at Cornell University were studying

[19] https://laborrights.org/in-the-news/six-cents-hour-1996-life-article.

[20] https://glassclothing.com/a-history-of-nikes-changing-attitude-to-sweatshops/.

[21] https://www.accenture.com/content/dam/accenture/final/a-com-migration/manual/r3/pdf/Accenture-CompetitiveAgility-GCPR-POV.pdf.

Fig. 3.3 A Sweatshop in India (Getty Images)

how emotions could travel via text messages to learn more on the gap between texting and talking. Facebook "outsourced" the study to some of these Cornell scholars, including Jeff Hancock, then Associate Professor of Communication at Cornell, with whom I had a conversation on the Facebook experiment in 2018 when he gave a talk that I organized at my university. The outcome of the study was a joint research paper with a Facebook researcher (Kramer et al., 2014) that reported the results of the experiment. In defense of the Cornell team, Facebook gave them already anonymized data to analyze. Once the experiment went public, Facebook users felt violated and betrayed. In the aftermath of the public disclosure of the experiment, Jeff Hancock told us in his public presentation that his inbox became flooded with hundreds of emails of people (Facebook users) asking whether they were involved in the experiment (note that the experiment was anonymous, users were selected randomly, and the "real" participants' names were never reidentified). The Facebook experiment received large (negative) media coverage. Bottom line: Facebook didn't lose traction, the scandal was soon forgotten, and the platform kept increasing its number of users.

Second, differently than Nike, which has and had many competitors, high-tech companies have historically operated in oligopoly (if not monopoly) regimes. Many my age who went to college when Internet was just blooming (in the mid-1990s) may recollect the dominant position of Microsoft Corporation in those years, which led to the US Department of Justice to file antitrust suit against it for "unlawful monopolizing computer software markets."[22] At the initial trial, the US District Court in Washington DC ruled that Microsoft's actions constituted unlawful monopolization under Sect. 3.2 of the Sherman Antitrust Act,[23] but a subsequent appeal partially overruled that judgment with the two parties later reaching a settlement in which Microsoft agreed to modify some of its business practices.[24] In my opinion, the Microsoft settlement was a landmark example of how high-tech companies would be allowed to operate in the Internet era, especially in America. The Federal Trade Commission (FTC) has prevented many mergers of large manufacturers and service companies to encourage competition (even during the laissez faire Reagan era) but has been extremely weak when called out to regulate high-tech companies. One very recent example (again) concerns Microsoft being able to acquire Activision, a gaming company, further consolidating the gaming industry (a US appeals court refuses FTC request to pause Microsoft deal for Activision).[25] While having a few companies building large-scale platforms might be seen as a good thing for economies of scale (e.g., Facebook can offer access to its platforms free of charge because they have nearly 3 billion users, that they can target with personalized ads), a company's dominant position doesn't give users a choice. While people can buy shoes from Nike as well as Adidas, Converse, Asics, and so on, social media users need to gather within the same platform because this is the whole point of social networks. Indeed, too many providers in a particular space would result in few users on each one—or users have to maintain different profiles in different social media platforms, something that we see not happening in the current social media scenario, where different social media are generally connotated by specific demographics such as age. This, however, creates a de facto monopoly of the social media company with the most users, which

[22] https://www.justice.gov/archive/atr/public/press_releases/1998/1764.htm.

[23] https://law.justia.com/cases/federal/district-courts/FSupp2/87/30/2307082/.

[24] https://www.wsj.com/articles/SB1000076767888491506.

[25] https://www.reuters.com/legal/us-appeals-court-refuses-ftc-request-pause-microsoft-deal-activision-2023-07-14/#:~:text=U.S.%20appeals%20court%20refuses%20FTC%20request%20to%20pause%20Microsoft%20deal%20for%20Activision,-Reuters&text=July%2014%20(Reuters)%20%2D%20A,O.

explains why in 2011 Facebook users, regrettably, have decided to stick with the platform even in the aftermath of its public scandals.

Network externalities play a substantial role in making people stick with online platforms. With AI, the role of network externalities plays a background role for users because, for instance, ChatGPT users don't interact with each other. But the extent to which an AI is being used by so many users matters to them because the more users, the better result due to how AI is trained. One good example of the relevance for AI platforms to gain and maintain a critical mass is Waze,[26] a popular GPS-equipped traffic navigator app that relies on users to provide route advice but, more importantly, to provide real-time traffic insights. Waze is reliable because many people use it, which means it can track, for example, their speed via GPS systems to inform other drivers about traffic and offer them alternative routes. Here, too, consolidation benefits users. The more drivers actively use Waze, the more feedback on traffic other drivers will create. But, once again, this poses a tradeoff between "good" outcomes of oligopoly and customer choices. To wit, high-tech companies trap users into network externalities benefits, which prevents newcomers from accessing the market and hooking customers in current apps, social media, etc. In Michael Porter's competition analysis, high-tech companies that offer AI services to the public would represent an industry with high barriers to entering the industry (Porter, 1985). This is quite a paradox for Internet companies as the online world has always been seen as one with very low barriers to new competitors, e.g., due to the lack of fixed costs, physical infrastructures, etc., but also because "big data" generates greater outcomes—thus, the better companies are at harvesting data, the more chances they have to succeed.

Automated Training

In contrast to manual training, which people perform by definition, computers that people program perform automated training. They are both attempts to classify the world with their own pros and cons. Manual training's pros mainly concern the fact that humans, who are smarter than computers, decide how to classify "things." But the shortcomings are substantial. As I illustrate in the previous section, labelers are far from

[26] https://www.waze.com.

perfect. They are poorly trained, put under a lot of stress, and are often exploited, which poses important DEI issues concerning how AI are built (even if, sadly, end users seem to care little to nothing about it). Automated training happens prevalently when AI systems are in production or, according to the algorithmic decision-making lifecycle that I describe in a recent paper (Marabelli et al., 2021a), when AI is *used in practice*. The most relevant threat to automated training, at least from the viewpoint of this book about DEI and classifications, is that this type of training relies on naturally biased historical data, especially given how different is the world in which we live today rather than even a decade ago (e.g., in terms of openness to LGBTQ + communities, attention paid to women in the workplace, and so on). One might argue that manual training is biased too. That's true. And that people are biased, racists, and so on. However, optimistically talking, historical data might well be worse than what people think today in the wake of the #metoo movement, George Floyd's murder, and other milestone events that have made communities reflect on the fact that, for instance, people's biological sex should be divorced from their gender. To wit, and still optimistically talking, what people think and do today with respect to, for example, race and judgment is (on average) better than what they would think and do decades ago. Even people who are careless of the social injustices that have perpetuated the past are aware of the issues at hand and know that now they can be held accountable. Even the most sexist man is aware that nowadays groping a woman in the workplace would lead to immediate firing, with legal consequences. Decades ago, this practice was not just accepted, but also praised and viewed as an exercise of masculinity and a display of men-power.

My point here is that as we can blame labelers for being shallow in training AI, they have a view of the society in which we live (which is not the 1980s). Automated training relies on historical data (so, something possibly (also) from the 1980s) along with Web data (scraped), which can be highly biased because the Web is not a diverse place. As I note earlier, even Wikipedia, considered by many as the gold standard to train automated system, is a very biased platform (Bender & Friedman, 2018). The so-called human in the loop is poorly used in AI automated training. All this poses important ethical and DEI issues of automated training. For one, think if Fox News, the most watched cable news network in America in 2023[27] with a debatable culture concerning women and race (a now fired anchor, Tucker Carlson, recently legitimized racism),[28] decided to use algorithms relying on their past hiring to hire new people. The underlying AI systems would identify the people at Fox who were most successfully (please, read = who made the most money) and would logically identify (and hire) White old men. Training data relying on datasets that perpetuate longstanding social injustices pose issues that are similar if not worse than settings in which AI is trained by incompetent humans. And with the advent of generative intelligence, AI systems are not data-hungry.

That AI is data-hungry is a well-known issue. The logic behind it is simple. AI doesn't really think and fells like humans do (even if someone thinks otherwise, like Blake Lemoine, a former Google engineer who, in 2022, suggested that Google's natural language models are sentient and show self-awareness, something that was later found to be unsubstantiated).[29] For instance, generative AI that produces text in response to a user's prompt (such as ChatGPT) is designed to probabilistically "guess" the next word based on the most likely word that can come next, allowing it to generate contextually accurate text. Facial recognition systems need extremely large samples of "faces" in order to pick the differences between them. One reason why these systems are weak in recognizing faces of people with dark skin relies on the relatively small samples of people of color being fed to them. Along with data privacy issues associated with AI scraping the Web (or Facebook using tags that people use on their photos) to train systems, we are heading toward a scenario where AI will run out of data. An international team of researchers including scholars form MIT's Computer Science and Artificial Intelligence Laboratory have recently conducted an analysis suggesting that the documents we have available in digital form (online and beyond, such as in library archives) might soon not be enough to train large models (Villalobos et al., 2022). This issue arises partly because language AI researchers

[27] https://www.forbes.com/sites/markjoyella/2023/02/01/fox-news-hits-23rd-consecutive-month-as-most-watched-in-cable-news-as-cnn-sees-gains-in-january/?sh=400978e15f59.

[28] https://www.theguardian.com/us-news/2023/jul/27/tucker-carlson-racism-book-biography.

[29] https://www.washingtonpost.com/technology/2022/06/11/google-ai-lamda-blake-lemoine/.

often categorize the data used for model training as either high quality or low quality, the latter including social media posts, blogs, comments that vastly outnumber those in the high-quality category. Researchers usually only train models using high-quality data because they want the models to reproduce it accurately. The possibility that researchers will start relying more on low-quality data is, therefore, concrete. And we all know that training AI systems with random text captured from websites such as Reddit and Quora will make AI systems more prone to use data that features discrimination, fake news, and more.

A contrasting argument related to large datasets is put forward by Dr. Andrew Ng, founder of DeepLearning.AI, Founder & CEO of Landing AI, General Partner at AI Fund, Chairman and Co-Founder of Coursera, and Adjunct Professor at Stanford University's Computer Science Department. He is considered one of the most competent authorities in the AI realm. He recently suggested that, in well-studied ML contexts, it is possible to shrink datasets to as few as 10,000 examples—"a sort of threshold where the engineer can basically look at every example and design it themselves and then make a decision" (Hao, 2021). Ng's intuition is supported by research suggesting that, in circumstances where outcomes are known, neural networks (AI computer systems modeled on the human brain and nervous system) can require only minimal training data. They call them shallow neural networks: small neural networks with only a few layers in which calculations take place (Vanian, 2021). One main limitation of Ng's approach is that small datasets can be used only in particular settings (well-studied contexts) while the new frontiers of generative AI involve exploring new settings rather than exploiting existing, well-known problems. Using small datasets can happen, for instance, by increasing data quality, by making sure that all the data used to train a model is "pertinent" to the model (often a lot of data used to train model is irrelevant, but discarding data is still a time-consuming process), and by ensuring that labeling is done comprehensively (e.g., in contexts where training data involves image-recognition systems as Ng explained in a recent *Harvard Business Review* article) (Ng, 2021). ChatGPT illustrates how AI is and will be required to answer to questions that are randomly generated by users on a nearly infinite number of topics. Whether AI models rely on big or shallow data, it is important to preserve the integrity of the data supply sources, something I am concerned about.

The training data supply chain uses humans or a natural phenomenon recorded by humans as its key (and, I would say, unique) data source. For instance, we are "walking data generators" (McAfee & Brynjolfsson, 2012) when we withdraw money at the ATM, wear a smartwatch, use our smartphone, purchase something online (or in a store with a credit card or a digital payment), or basically when we do anything in our life unless we purposely aim to avoid being tracked and recorded. Note that doing so might be near to impossible but also risky. In the era of surveillance capitalism, one who suddenly stops producing digital traces (as we academic call data that our online activities produce or that digital devices otherwise record—something that the press calls data exhaust) might rise red flags (e.g., with the IRS (internal revenue service) or the police). This happens because we are accustomed to being able to Google people to check their legitimacy. I experienced this feeling firsthand. I recently was contacted by someone who said she was a retired professor at a local university, which she used as an explanation for why she didn't have a webpage at the university. She didn't have a LinkedIn profile. She didn't have any social media presence, a website, or anything else. Her email, however, came from the university she said she was still affiliated with. She asked me to provide support to professors from a Global South country who wanted to attend a conference that I co-organized, which meant I could assign waivers warranting attendance to the conference free of charge to eligible people (precisely people form the Global South).

Because it is easy to forge emails addresses, I initially declined to provide waivers. I contacted others in the conference organization committee, and they all agreed that it must have been a scam attempt. Quoting an email from a colleague: "this whole thing stinks." Yet, for a reason still unclear in my mind, I decided to perform an additional check and contacted the department chair of the university this person said she was affiliated. It turned out that this person was a rea *emerita* professor, someone eclectic enough to have no online presence. I provided the waivers and learned an important lesson. We are too much used to believe that Google rules, and that that it is not OK to conform to common practices, such as having an online presence. But this is also an example of the consequences of not having one. It is a tradeoff. But, unfortunately, to find a job, access networks, and even access credit, it is better we are online; in fact, as I explain in Chapter 1, banks make decision on whether to give us a loan (also) based on our social media links, and not having online presence may prevent someone from getting one. While I am not saying that it is OK to be forced to have an online presence, I am just recognizing how the world spins. Turns out that, in the era of data walking generators, we are the primary source of the training data supply chain. The push to be online for social reasons is a blessing for data-hungry systems because we are forced to continue producing digital traces or else run the risk of raising red flags with friends, colleagues, perspective recruiters, and even with law enforcement. Other primary data sources besides human beings include wind speed, atmospheric pressure, currents, and temperature (e.g., in weather-forecasting models) or other natural data collected during experiments of various kind (car crash tests, chemical reactions, etc.).

The bottom line is that the starting point of training data supply chains is primary data—observations whether about our behavior or natural phenomena. The rest the supply chain unfolds as follows: the data is collected by digital devices (so it is *digitalized*), funneled into processing systems, *classified*, compared with existing data, cleaned (if needed), and then supplied to an AI system. Let's remember that all data comes from an analogic system—our world. One simple example will shed more light on the training data supply chain. Consider a person who takes a selfie with a smartphone (here the digitalization process takes place) and uploads it to Facebook. The Facebook platform (processing system) records the data (image) as belonging to a certain user and recognizes that it is an image of a face (better if the person performs the labeling process on their own). Furthermore, the processing system classifies the face as male/female (it is hard to think that a classification system would, on its own, recognize non-binary individuals, and it would be worrisome if it happened automatically based on how someone appears in a photo). Additional classification attributes might include skin color or other profiling characteristics. In addition to this process, manual classification might intervene as needed. Out of billions of photos, many will be discarded because, with low definition, they feature too much ambiguity. This would be the data-cleaning step. Once the (digitalized) data is cleaned and classified, it is given to an AI system that needs to be trained. Data such as text in a Wikipedia page is still primary data because humans generated it. A digitalized version of a book is primary data as well. The supply chain of natural phenomena works pretty much in the same way with the difference that data is generally more objectively classified; the speed of winds is measured in miles/hour or feet/second, while it is hard to measure ethnicity based on skin tone.

Institutional review boards (IRB) may represent the problem *and* the solution to training data supply chain being disrupted, if secondary data is vetted (by IRBs) seriously. In social science research, we define secondary data as data that has been collected by someone else (so already processed rather than raw data), and this data is generally available to the general public, even if secondary data is sometimes "for sale" by data brokers or held in government databases. Economists often purchase data from central banks or private data-harvesting companies that conduct specific studies to collect data that can be sold to academics for research purposes. Research ethics principles recommend that scholars should treat secondary data with the same scrutiny as primary data. For instance, when downloading medical data from an online public database, one should ask questions such as: was the data properly anonymized? Is it legit, legal, and ethical to use this (pubic) data for research? Does the data come from a reliable source? And so on.

In sum, secondary data does not warrant integrity. Even if the data is provided by entities with a reputation, such as the government. For instance, research on census data in the US is controversial because the methodology to acquire this data is controversial in the first place as a Yale University report has explained.[30] In addition, some institutions (wrongly) consider secondary data as safe by definition in issuing IRB clearance. In the US, IRB review is a process required when researchers conduct research involving human beings, and it ensures that human subjects are not harmed by being study participants since research that harmed people would be unethical according to the Collaborative Institutional Training Initiative (CIT), an initiative aimed at "serving the training needs of colleges and universities, healthcare institutions, technology and research organizations, and governmental agencies, as they foster integrity and professional advancement of their learners."[31]

The process is the following: a researcher who wants to study individuals (e.g., interview people) needs to provide to their university's IRB with details on the people being interviewed, information collected, questions asked, and so on. Then, the IRB either approves or denies the study. Approval would mean that the research is worth being conducted or that

[30] https://onlineexhibits.library.yale.edu/s/Census/page/controversies-about-U-S-census.

[31] https://about.citiprogram.org.

the benefits from the research overcome the potential risks. In social sciences, we aim to conduct research with minimal to no risks (Marabelli & Markus, 2017). But, in other contexts, it is possible that an IRB approves research that involves some risks because the potential outcomes would be so meaningful that some risks become acceptable. For instance, during the COVID-19 pandemic, some research projects involved intentionally inoculating the COVID-19 virus into healthy individuals to study new drugs (Sidik, 2023). These kinds of studies are clearly very risky, but the IRB has evaluated that the outcomes could overcome these risks.

Therefore, the fact that some IRBs consider secondary data as "safe" by definition is problematic. On the one hand, secondary data might have been collected in unethical ways (e.g., by conducting covert research without proper debriefing). This means that data was collected without the study participants knowing they were data sources. This happens relatively often in sociology and psychology when informing people that they are part of a research would condition their behaviors. But, after the data is collected in a covert manner, the researchers are required to disclose the data-collection process, explain the research to these people and ask them if the information collected can be included in the dataset. In other words, this is an ex post administration of an informed consent form. If the dataset doesn't detail how covert research was conducted, this dataset would be suspicious (and an IRB should precisely study these cases). But if secondary data is not required to obtain IRB clearance, we risk conducting research on data collected unethically. The example of the Facebook mood experiment is a good one to showcase how research can be conducted in questionable ways when IRB clearance is not required for secondary data. The Cornell scholars were given Facebook users data and, thus, considered it secondary data. They were able to analyze data on manipulated feeds (unethical). Yet, because the Cornell scholars did not collect this data in the first place, they were able to conduct their analysis. As I explain in an article with colleague M. Lynne Markus, the Facebook experiment was undertaken with neither IRB approval nor users' consent (or post hoc debriefing)—never mind that university-based researchers participated in the data analysis. In analyzing the Facebook experiment, Grimmelmann (2015) concluded that social media experiments afford opportunities for "IRB laundering." Facebook could capitalize on loopholes in the current system to regulate research to circumvent human subject protections (Marabelli & Markus, 2017).

A similar and more general example of lack of IRB vetting concerns big data. Some have argued that, because big data is *big*, it is impossible to reidentify people (the process by which a data point is linked back to the data creator, such as a blood pressure reading and a patient). It is also impossible to administer informed consent form. For these reasons, some have argued that the IRB process should be waived when one conducts this kind of research. This idea is problematic because (1) people can be identified from big data—Latanya Sweeney (2016) provides several argument in favor of this scenario and (2) reidentification is not the only way to create harms to people. Clearly, a dataset on medical data of people with HIV that turned out not to be properly anonymized would represent an extremely important privacy breach because HIV still carries social stigma and can lead to discrimination and even prevent people from accessing some jobs. But simply manipulating online posts of users (as per what happened in the Facebook experiment) is highly unethical as it generated unpleasant feelings among the Facebook community.

Here I raise three questions. I have a short answer for the first two questions, while the third question deserves broader discussion. First, is data training for AI considered research? It probably should be because it involves data harvesting to generate an outcome. Second, should AI research looking for training data be IRB exempt? Probably not, since collecting data might jeopardize people's privacy (cf. when Italy blocked ChatGPT for a month in April 2023).[32] Third, what is considered secondary data for AI? If we apply the above rules for research conducted by humans to AI, we end up realizing that, for an AI system, secondary data is data collected, analyzed, and classified by other AI systems. While this as first might seem something very dystopian, it is actually already happening and is a phenomenon that, in my view, will grow in magnitude and will become increasingly problematic. Generative AI can produce texts and images from prompts and in a systematical manner, which means that it can (and does) contribute to social media content much more effectively than traditional bots. The 2024 US presidential election is already believed to be in jeopardy due to the potential influence that generative AI might have on it (e.g., via influencing online discussion, disseminating disinformation, generating fake news, or just spreading propaganda for this or that party or candidate).

[32] https://www.nytimes.com/2023/03/31/technology/chatgpt-italy-ban.html.

Given that training data is mostly harvested from online content, it is easy to predict that AI Cannibalism Disease (AICD)—the scenario when AI, which is data-hungry, is fed with its own data—will be a widespread phenomenon in the near future when AI is fed with its own data. Pessimistic (in my opinion) forecasts such as that of Nina Schick from Yahoo[33] assess that, by 2025, 90% of online content could be generated by generative AI. I don't think that these figures are realistic. Yet, I wouldn't be surprised if 20% of the online content was based on AI in a few years. That's a lot. Given that AI is trained mainly with "legit" online data (e.g., Wikipedia) rather than data from social media, forums, discussion boards, and so on, it is possible that ~ 5% of training data would "contract" AICD. This, to me, is enough to disrupt the data training supply chain. The effects of AICD are unknown, but we already know that AI can be racist (or, better, can act in ways that reflect people's racism). What if awareness of social justice, respect for diversity, and praise for minorities grows into populations' cultures while AI systems act otherwise? People are "fed" with current data as I previously said. Even people not particularly versed in DEI matters know that, in 2023, they can no longer use certain (derogatory) language or act in ways that might have been acceptable even 10 years ago. But human beings are trained with current data because they experience current society (they read newspapers, watch television, etc.). AI systems are trained with current *and* historical data can paradoxically act in ways that are much more racist than people. If AI starts populating the Web even if minimally with more racist content than what human beings would post, future AI might become one where these systems will become much more biased against minorities than they already are.

Some (including Gartner) have predicted that, in the future, we might use "synthetic data" (data produced by AI to train specific models) partly for training processes.[34] However, planning to generate synthetic datasets is different than doing so not willingly. Synthetic data might be needed given the need to correct biases embedded in Web data. But, drawing on Ng's idea, synthetic data, along with strategies and designs aimed at using small neural networks have the potential to address another DEI

[33] https://news.yahoo.com/90-of-online-content-could-be-generated-by-ai-by-2025-expert-says-201023872.html.

[34] https://www.wsj.com/articles/fake-it-to-make-it-companies-beef-up-ai-models-with-synthetic-data-11627032601.

issue affecting training data, which concerns the remarkable amount of energy required to perform training processes.

AI and the Environment

One last significant issue with training AI systems that is often overlooked is its environmental impact. Data centers and data transmission networks accounted for 2% of total greenhouse gas (GHG) emissions in 2022.[35] That could be surprising for some since going green implies going digital. While this is true for some of our activities, it is inaccurate for all digital activities. For instance, scraping the Web to collect training data, streaming video surveillance (from Amazon warehouses), and video reviewing (workers behind Amazon systems) consume electrical energy (personal computers, servers in the data centers, etc.). Global Internet traffic has grown exponentially over the past 30 years, which has contributed to greater energy consumption and more environmental consequences. For instance, in 2022, global Internet traffic was expected to amount to 153,000 GB of traffic per second (4.8 zettabytes) with personal data constituting a significant part of the total data transfer.[36] More specifically, if we refer to the use of AI, just to give an example, training a single AI model can be compared to that of several jets carrying thousands of passengers for a long-haul trip (Strubell et al., 2019).

As I argue in a recent article that, among other things, discusses the relationship between training AI models and global warming (Marabelli & Newell, 2023), it is important to reflect on whether it is worth building such sophisticated systems that, as we have seen, carry several potential negative consequences for the very people they are designed for (end users). For instance, a 2017 United Nation study systematically analyzed climate change with respect to social inequality and concluded that "inequality causes the disadvantaged groups to suffer disproportionately from the adverse effects of climate change, resulting in greater subsequent inequality" (Islam & Winkler, 2017, p. 1). The problem is that—assuming we all agree that global warming is actually happening

[35] https://www.forbes.com/sites/forbestechcouncil/2021/05/03/renewable-energy-alone-cant-address-data-centers-adverse-environmental-impact/?sh=95bb0985ddc9.

[36] https://wdr2021.worldbank.org/stories/crossing-borders/.

and that a very fast pace is in part due to our irresponsible use of our planet's resources—the consequences of global warming are not equal. India (nearly 18% of the world population) generates 3% of the world's pollution but pays a high price because of global warming with its temperature having risen to 110F (37.3 C) in June and July 2022.[37]

CNN reported that, in June 2023, temperatures in India reached the "limits of 'survivability'."[38] In fact, during that month, in Uttar Pradesh, a state in northern India, the temperature reached 116F (47 C), which affected 220 million people and caused nearly 50 deaths. The same happened in the US where the southern states received several heat waves during its summer in 2023. For instance, Phoenix, AZ, has reached temperatures over 109F (43 C) for the whole month of July. But the difference between Arizona and Uttar Pradesh is substantial. Most residents in Arizona live in Phoenix where most people are well prepared for hot temperature and have air conditioning systems (between 95 and 99% of homes in the city have air conditioning systems).[39] This means two things. First, people can struggle with being able to work out outdoors daytime and may have to pay higher electricity bills. Second, and more consequentially, the heavy use of AC in houses challenges the state's electric grid, which requires it to make more investments in electric energy, adds to our demand of earth resources, and potentially harms the environment. In other words, and to generalize, more heat generated by the Global North leads residents to fight high temperature by using resources that contribute to global warming even more. In Uttar Pradesh, India, very few people have air conditioning systems.[40] Countries and populations that contribute the least to global warming pay the highest price. Environmental issues provide one of the clearest pieces of evidence of the social injustice produced by AI systems.

Data centers and their proliferation represent another important environmental issue. These facilities power AI and house the most

[37] https://www.technologyreview.com/2022/07/05/1055436/download-india-deadly-heatwaves-climate-change-carbon-removal/.

[38] https://www.cnn.com/2023/06/26/india/india-heatwave-extreme-weather-rain-intl-hnk/index.html.

[39] https://www.theguardian.com/us-news/2022/jan/27/phoenix-arizona-hottest-city-cooling-technologies.

[40] https://www.npr.org/sections/goatsandsoda/2022/08/02/1114354904/opinion-life-hacks-from-india-on-how-to-stay-cool-without-an-air-conditioner.

advanced computational systems that perform training processes. They need powerful ways to cool down servers, which reach high temperatures while processing data and information. One solution, pursued in the past by Facebook (now Meta), was to move data centers to countries with low average temperatures, such as Sweden.[41] This helps cooling down servers using natural resources. But this is not always possible for reasons spanning from the availability of locations where companies can build large facilities to the challenges to connect such facilities to the Internet at high speed. Also, generally very cold locations are in remote, difficult-to-reach sites with weak power grids. I also argue that attempts to disseminate data centers around the world (especially including remote locations) are themselves detrimental to the environment. Large-scale settlements of data centers in locations so far untouched by industrialization would simply spread industrialization with the obvious risk to unsettle flora and animal life.

Other solutions, such as that being pursued by Meta in Spain, are equally if not more problematic. In Spain, water has been famously a scarce resource. Spain is located in southern Europe and is prone to very high temperature in its summer with only mild temperatures in winter and scarce rain precipitation. In 2022, Spain struggled with drought to the point that local authorities recommended residents to be parsimonious with water and to not water their yards. According to a Bloomberg report,[42] the region of Castilla La Mancha, which produces a quarter of all Spanish grain, was expected to lose up to 90% of the 2023 harvest. Nevertheless, in Talavera de la Reina, a city in central Spain famous for producing decorative ceramics, Meta is planning to build a US$1.1 billion data center, which is expected to use around 176 million gallons of water/year for cooling. But cutting-edge innovations are promising. For instance, Heata,[43] a British startup, recently developed technologies using heat generated by computers to provide free hot water. While this solution doesn't address servers' demand to be cooled off, it at least mitigates the environmental impact of cooling demand of data centers supporting cloud computing.

[41] https://www.theverge.com/2016/9/29/13103982/facebook-arctic-data-center-sweden-photos.

[42] https://www.bloomberg.com/news/articles/2023-07-26/extreme-heat-drought-drive-opposition-to-ai-data-centers.

[43] https://www.heata.co.

Nevertheless, data centers hosting cloud computing services that harvest and process data, provide storage to subscribes, and run large-scale AI training systems represent an environmental hazard. While it is true that AI systems are now employed to accurately predict weather forecast,[44] if we keep using scarce resources to power these systems, AI will be simply able to predict the damages it is causing. This doesn't mean that we should give up AI because of global warming—training AI models contributes (so far) a minimal part of the global warming. Also, training AI systems will make them better in working out solutions concerning existing environmental problems. But it is important to use AI responsibly and make sure that there is a net positive between what AI can do to predict weather condition, for instance (along with other AI potential, e.g., to help design environmental-friendly innovations) and what natural resources need to be mobilized to run and train AI models.

Smart cities are an example of where AI can make up its infamous reputation as being detrimental to the environment because of the high computational demand to train its models. For instance, in smart cities, properly trained AI systems will be able to predict and manage traffic effectively. Waze,[45] an AI-powered community-driven initiative that I mention earlier in this chapter, can already reroute drivers to less congested routes and, thus, minimize people's driving and pollution. In smart cities, apps such as Waze could be centralized and employed to communicate with all citizens (e.g., using push phone notifications to advice *any* drivers about traffic issues). In 2021, *AI Magazine* illustrated ten ways in which AI can be effectively employed to support smart cities,[46] such as improving security via surveillance cameras, controlling pollution with meters, optimizing parking, tracking public transportation usage to meet but never exceed demand, using robots to manage garbage and recycling processes, and so on.

Clearly, building a smart city comes at a price. This price is literally a cost, which is required to "rewire" entire cities and equip them with a combination of connected devices (e.g., the Internet of things) and AI. An obvious outcome is that these cities will be appealing by the

[44] https://www.technologyreview.com/2023/07/11/1076067/weather-forecasting-is-having-an-ai-moment.

[45] https://www.waze.com.

[46] https://aimagazine.com/top10/10-ways-ai-can-be-used-smart-cities.

most, which means that, for the simple demand/offer principle, living in smart cities will be a privilege for the few. When considering global warming, I note above that populations that contribute the least to it pay the hardest consequences. In the case of smart cities, I tend to draw similar conclusions. People who can afford to live in big cities will be able to benefit from the advantages of being managed by AI, which involves reduced pollution, having better systems to face extreme weather conditions, lower prices (for instance, with energy efficiency housing and better insulated), and an overall higher quality of life (for instance, with smart traffic and parking, people will spend less time driving and more time at home with their families). In contrast, cities that are "too poor" to be turned into smart cities will continue to be crowded with people at the margin who cannot afford better housing settings. In America, housing redlining—lenders' ability to deny mortgages to Black people in expensive neighborhoods mainly that White people mainly inhabit—occurred de jure until the 1968 Jones vs. Mayer Supreme Court ruling,[47] but structural mechanisms involving state and government investments could still de facto perpetuate it in perspective smart cities, something already discussed with respect of critical race theory (CRT).

But, even in smart cities, residential racial segregation can easily happen. For instance, San Francisco, CA, is believed to be one of the most "advanced" smart cities in America,[48] yet it also one of the cities with the largest percentage of permanent homeless people. California, despite being the fifth largest economy in the world, hosts half of the unsheltered homeless population in the country[49] with a significant share of this population concentrated in San Francisco.[50] I think that San Francisco is the typical example of "Poverty, by America"—to quote the title of a recent book of Matthew Desmond (Desmond, 2023)—which highlights the gaps between the rich and the poor that the American way of conceiving capitalism has produced in the last century. With the segregation-by-neighborhood practices in San Francisco, smart cities

[47] https://supreme.justia.com/cases/federal/us/392/409/.

[48] https://www.thesmartcityjournal.com/en/cities/smart-city-san-francisco#:~:text=Smart%20City%20Platform%3A%20San%20Francisco,security%20threats%20in%20real%20time.

[49] https://www.auditor.ca.gov/issues/briefs/housing-and-homelessness.

[50] https://www.mckinsey.com/industries/public-sector/our-insights/the-ongoing-crisis-of-homelessness-in-the-bay-area-whats-working-whats-not.

become smart only for rich people with little regard for people at the margin. In fact, investments in AI-powered innovation to make cities such as San Francisco more livable for the rich risk doing so at the expense of people who cannot take advantage of such investments. For instance, how would a "smart" traffic system that minimizes driving commute and optimizes parking benefit someone who doesn't own a car (and a house)?

The international context is one such thing we need to consider when reflecting on the environmental impact of emerging technologies and, in particular, what it takes to train AI systems as I illustrate in this chapter. My view of the worldwide scenario of technology development involves the tradeoff between uncontrolled innovation vs. respect for people (privacy, freedom, equity, inclusion), and different countries have taken different approaches. As I've already reported, America, consistent with its obsession to reward who is already in power (while leaving behind who is already struggling), is all for unleashing innovations and using people to test them. If innovations discriminate, someone will "patch" them and make them more innocuous. But the damage is already done. Related to smart cities, the recovery is difficult, if not impossible, as I explain in Chapter 1 when describing redlining policies until the Fair Housing Act in 1968. The problem is that generations of Black Americans are currently paying the consequences of not having been able to purchase houses prior to 1968 (and after via racist lending practices) and, thus, not be able (now) to leave homes to their children and grandchildren. According to a 2016 PEW report, Black and Hispanic households are about twice as likely as White households to rent their homes. Furthermore, 58% of Black household individuals and 54% of Hispanic household individuals rented their homes compared with 28% of Whites.[51] AI contributes to the perpetuation of social injustice with landlords using it to advertise homes and apartments, using tenant selection systems fueled by AI and ads based on demographic microtargeting to leave out Black communities, thereby supporting discrimination of minorities, according to the 2019 Fair Housing Trends Report.[52]

Smart cities, in America, are and will be just for the rich—a consequence of America as "the land of the free." It is true that America is the

[51] https://www.pewresearch.org/short-reads/2017/07/19/more-U-S-households-are-renting-than-at-any-point-in-50-years/.

[52] https://nationalfairhousing.org/wp-content/uploads/2019/10/2019-Trends-Report.pdf.

land of the free…but you need to be able to afford it. Given this past, how training AI takes place is paramount given that, for example, historical data on lending is clearly skewed because de jure (prior to 1968) and de facto (until now) practices populate datasets. This is an example of how, in America, structural racism is mangled with how AI systems work and the environment when considering smart cities and existing barriers to access them by people at the margin.

In Europe, however, the music is different. The EU AI Act ensures that AI systems are overseen by people and are safe, transparent, traceable, non-discriminatory, and *environmentally friendly*.[53] Housing (and homeless) is an issue in Europe too, especially in large countries such as Germany.[54] Yet, the EU has strong plans associated with the 2021 European Green Deal (EGD). A landmark document focusing on AI and the EGD spells out[55]:

> The European Green Deal pursues ambitious environmental goals which require a green transformation of many sectors of society. The transformative potential of Artificial Intelligence (AI) to contribute to the achievement of the goals of a green transition have been increasingly and prominently highlighted. At the same time, digital technologies such as AI considerably increase energy and resource consumption and create risks of adverse environmental effects. This ambiguous picture illustrates that political and regulatory action is necessary to channel the potential of AI towards the goals of the European Green Deal. Tailoring and implementing effective AI policies presuppose a solid understanding of the socio-technical mechanisms that may lead to desired or undesired consequences of AI. This report aims to contribute to the development of such an understanding.

This doesn't mean that the EU is and will be at the forefront of green AI, but initiatives such as the EGD clearly set the stage for technology innovation that accounts for the relevance of environmental issues. But that's a first step in the right direction. Other AI economic powers such as China haven't explicitly acknowledged AI's threats to the environment. Yet, even

[53] https://www.europarl.europa.eu/news/en/press-room/20230505IPR84904/ai-act-A-step-closer-to-the-first-rules-on-artificial-intelligence.

[54] https://landgeist.com/2021/08/24/homelessness-in-europe/.

[55] https://www.europarl.europa.eu/RegData/etudes/STUD/2021/662906/IPOL_STU(2021)662906_EN.pdf.

if not specifically focusing on AI, the second largest world economy has taken important steps to address its carbon emissions according to the World Economic Forum.[56] China, similar to the US, is heavily working on generative AI along with powerful regulations concerning privacy and use.[57] While it is not clear how this will fit into China's grand vision of becoming "greener",[58] we need to monitor closely how the country will meet environmental needs concerning AI training and computation.

Conclusions

In this chapter, I outline what I believe are the main issues associated with training AI models. My main concerns relate to the exploitation of marginalized populations, especially in the Global South (manual training), and how data is harvested from online sources. The future of training AI poses many challenges, including the availability of data and the extent to which training systems will impact the environment. Sensitivity to environmental issues related to AI is very country specific as are all environmental issues more generally. Exploitation of people at the margin along with automated collection of biased data create climate-related challenges that are unevenly distributed across the world. Or, to put it differently, global warming affects every region of the world, yet some are more prepared (or have more resources) to face it. 115F (46 C) in Texas is different than 100F (37.7 C)in Pakistan as the latter doesn't have widespread diffusion of air conditioning systems, workers operate mostly outdoors, and, ironically, Pakistan releases much less CO^2 per capita than the state of Texas (even though Pakistan has 225 million people country and Texas has 29.5 million people tenth). In fact, Pakistan emits 0.85 metric tons of CO_2 per capita, while Texas emits 622.4 metric tons per capita. I leave any further comments on this data to the reader.

[56] https://www.weforum.org/agenda/2018/04/china-is-going-green-here-S-how/.

[57] http://www.cac.gov.cn/2023-07/13/c_1690898327029107.htm (this article is in Mandarin Chinese and was translated and vetted by a native speaker).

[58] https://www.china-briefing.com/news/what-is-chinas-green-and-low-carbon-plan-and-why-is-it-relevant-to-foreign-investors/.

References

Bender, E. M., & Friedman, B. (2018). Data Statements for Natural Language Processing: Toward Mitigating System Bias and Enabling Better Science. *Transactions of the Association for Computational Linguistics, 6*, 587–604.

Bender, E. M., Gebru, T., McMillan-Major, A., & Shmitchell, S. (2021). On the Dangers of Stochastic Parrots: Can Language Models Be Too Big? *Proceedings of the 2021 ACM Conference on Fairness, Accountability, and Transparency*, 610–623.

Desmond, M. (2023). *Poverty, by America*. Random House Large Print.

Grimmelmann, J. (2015). The Law and Ethics of Experiments on Social Media Users. *Colorado Technology Law Journal, 13*, 219–272.

Hao, K. (2021). Andrew Ng: Forget About Building an AI-First Business. Start with a Mission. *MIT Technology Review*. https://www.technologyreview.com/2021/03/26/1021258/ai-pioneer-andrew-ng-machine-learning-business. Accessed on April 10, 2021mi.

Islam, S. N. and J. Winkler (2017). "Climate Change and Social Inequality." *Department of Economic & Social Affairs* Working Paper No. 152, United Nations, available at https://www.un.org/esa/desa/papers/2017/wp152_2017.pdf, accessed on December 20, 2023

Kramer, A. D., Guillory, J. E., & Hancock, J. T. (2014). Experimental Evidence of Massive-Scale Emotional Contagion through Social Networks. *Proceedings of the National Academy of Sciences*, 8788–8790.

Marabelli, M., & Markus, M. L. (2017). Researching Big Data Research: Ethical Implications for Is Scholars. *Americas Conference of Information Systems (AMCIS).*

Marabelli, M., & Newell, S. (2023). Responsibly Strategizing with the Metaverse: Business Implications and DEI Opportunities and Challenges. *The Journal of Strategic Information Systems, 32*(2), 1–16.

Marabelli, M., Newell, S., & Handunge, V. (2021a). The Lifecycle of Algorithmic Decision-Making Systems: Organizational Choices and Ethical Challenges. *Journal of Strategic Information Systems, 30*, 1–15.

Marabelli, M., Vaast, E., & Li, L. (2021b). Preventing Digital Scars of Covid-19. *European Journal of Information Systems, 30*(2), 176–192.

McAfee, A., & Brynjolfsson, E. (2012). Big Data: The Management Revolution. *Harvard Business Review, 90*(10), 61–67.

Ng, A. (2021, June 29). AI Doesn't Have to Be Too Complicated or Expensive for Your Business. *Harvard Business Review*: online article. https://hbr.org/2021/07/ai-doesnt-have-to-be-too-complicated-or-expensive-for-your-business

Noble, S. U. (2018). *Algorithms of Oppression, in Algorithms of Oppression*. New York University Press.

Obermeyer, Z., Powers, B., Vogeli, C., & Mullainathan, S. (2019). Dissecting Racial Bias in an Algorithm Used to Manage the Health of Populations. *Science, 366*(6464), 447–453.

Perez, C. C. (2019). *Invisible Women: Data Bias in a World Designed for Men*. Abrams.

Phillips-Fein, K. (2009). *Invisible Hands: The Businessmen's Crusade against the New Deal*. W. W. Norton.

Porter, M. E. (1985). Competitive Strategy: Creating and Sustaining Superior Performance. *The Free, New York*.

Schwemmer, C., Knight, C., Bello-Pardo, E. D., Oklobdzija, S., Schoonvelde, M., & Lockhart, J. W. (2020). Diagnosing Gender Bias in Image Recognition Systems. *Socius, 6*, 1–17.

Sidik, S. (2023). 'Bold'study That Gave People Covid Reveals 'Supershedder'phenomenon. *Nature, 618*(7967), 892–893.

Smith, M., & Miller, S. (2021, April 13). The Ethical Application of Biometric Facial Recognition Technology. *AI & Society* (published), pp. 1–9.

Strubell, E., Ganesh, A., & McCallum, A. (2019). Energy and Policy Considerations for Deep Learning in Nlp. *arXiv preprint*. https://arxiv.org/pdf/1906.02243.pdf%22%3EWachstum

Sweeney, L. (2016). The Importance of Data Privacy and the Impact of Technology. The Grace Hopper Celebration of Women in Computing (GHC) Conference. https://www.youtube.com/watch?v=UBzP0NouiGo

Vanian, J. (2021). The Benefits of 'Shallow' Artificial Intelligence. *Fortune*. https://fortune.com/2021/05/04/the-benefits-of-shallow-artificial-intelligence/

Villalobos, P., Sevilla, J., Heim, L., Besiroglu, T., Hobbhahn, M., & Ho, A. (2022). Will We Run out of Data? An Analysis of the Limits of Scaling Datasets in Machine Learning. arXiv preprint. https://arxiv.org/pdf/2211.04325.pdf. Accessed on December 8, 2022.

CHAPTER 4

Discipline, Punish … and Workarounds

AI's ability to extract, classify, and process data on individuals and groups, as I explain in the previous chapters, necessarily leads to various forms of control. For instance, in Chapter 2, I discuss how Uber drivers can feel uncomfortable because the app manages their rides and monitors the minutiae of their work-life. In this chapter, I further expand on the Uber driver example (among many others) and rely on two studies that I recently conducted with data collected in the US and the UK. Algorithms powering AI represent a source of oppression as Safiya Noble (Noble, 2018) broadly highlights in her book. Yet, Uber drivers, along with many other exploited (and surveilled) workers, are very resourceful and can react to the "power of algorithms" in very creative ways—a phenomenon that (Kellogg et al., 2020) call *algoactivism*. Social media users behave in a similar way. Their data is used, in real time, to sell personalized ads. Social media users are not just stripped of their privacy. They are also nudged, which means that they might be led to buy what they don't need. Yet, social media influencers can take advantage of algorithms (because they understand how they work even if partially). It is worth noting that nudging practices don't happen only online. For instance, when grocery store marketers place amenities (candies, treats, gums, etc.) on shelves that people in line to check out can see, they are nudging them by exploiting their boredom while they wait for service. But social media

M. Marabelli, *AI, Ethics, and Discrimination in Business*, Palgrave Studies in Equity, Diversity, Inclusion, and Indigenization in Business,
https://doi.org/10.1007/978-3-031-53919-0_4

nudging is different because it is targeted and relies on data collected via online surveillance. There's also what I call *second degree* nudging, which happens with user data traded by brokers in ways that are unclear and might be illegal. For instance (anectodical evidence, yet astonishing), in June 2023, my wife and I went to a well-known pastry shop in Boston's North End neighborhood. The shop is cash only. We purchased three items after dining in a nearby restaurant and took the pastries home. The day after, my wife's Facebook featured ads of pastry shops with many ads for the pastry shop we visited the day before. As we didn't use any of our devices to Google the shop, we were wondering if, by paying with a credit card at the restaurant across the street, we gave data brokers some leads. Not having smart (listening) devices at home, my wife wondered if our phones could listen to our conversations, which is possible though unlikely. My best guess was that we were victims of data triangulation. Various digital traces such as smartphone location data (restaurant and pastry shop), credit card records (restaurant), and AI inferences generating patterns (others' habits of purchasing pastries after a meal in that neighborhood at a certain time of the day) led to the Facebook ads we saw.

The pastry shop story concerns microsurveillance and digital nudging and involves near-illegal data exchange across platforms. Furthermore, it is difficult to contrast it with simple practices such as disabling location tracking apps. Other types of surveillance are more evident to users and can be effectively contrasted—examples concern applying for a small loan and knowing how much it matters to have a "good" FICO credit score (see Chapter 1 and access to credit in the US). Then there are workplace-related types of more or less invasive types of control all connected to people analytics (see Chapter 2). For instance, in healthcare, control is not based on video surveillance (e.g., as in Amazon warehouses). In healthcare, algorithmic-based control revolves around data generated by enterprise systems such as EPIC.[1] Clinicians and staff are the data generators. Algorithms work on that data and offer administrators insights that can lead to performance management and suggest actions (process improvement but also restaffing and, in some case, resizing), yet, it is another form of control. The difference between healthcare surveillance and other types of workplace surveillance is that, in healthcare,

[1] https://www.epic.com.

generally clinicians and staff members have a clear idea of how they should act to perform well. The data is still collected on their everyday work practices, but patient outcome variables are transparent. This type of control happening in professional bureaucracies—organizations where performance management focuses on outputs rather than processes—leads to employee dynamics that are totally different than those happening in machine bureaucracies such as Amazon.

The examples above, which I unpack in the reminder of the chapter, suggest that algorithmic control varies across settings and is not a black or white concept. Algorithms are not always a blackbox, nor are they always transparent. Their design and applications represent a continuum with contexts where their behavior is predictable (credit scores for small loans), somewhat ambiguous (Uber drivers, social media, healthcare), or totally blackboxed (second degree nudging, mortgage applications where algorithms rely on social media). In addition, the technical features built around algorithms and AI can be fully blackboxed, fully open, or mixed. For instance, some algorithms are designed in a transparent way, which means that input–output variables are part of the blueprint of an algorithm. Programmers (and management, who makes decisions) can understand why some inputs lead to some outputs. An example is the performance management designed for Uber drivers. While the drivers are not sure how their behaviors will be evaluated, the variables are specifically designed by the programmers to offer executives with a decision-making dashboard. This leads to obvious information asymmetry between drivers and management, an important issue that I examine later in this chapter. Hiring software, which I discuss in Chapter 2, are what I call a mix. Here, algorithms are rather transparent for the designers and for the managers who use outcome variables to make hiring decisions. However, the many variables used to screen resumes are often collapsed into yes/no outcomes rather than a performance dashboard. At Uber, management needs dashboards to evaluate drivers; some variables might showcase that a driver is always on time but not kind with customers, which would lead to suggestions for improvement. However, hiring software is created for the sole purpose of discarding resumes. Therefore, the outcome variables, albeit transparent from a technical standpoint, are not transparent to decision-makers who simply find a pile of resumes on their desk—(the ones that the algorithm did not discard). Similar to Uber, in the case of hiring software, end users have only some clues on how the algorithms that they interact with work.

Other algorithms, used to process large amount of data, are much more complex than Uber and hiring software algorithms. In these instances, even programmers have no clue of the how and why some algorithms lead to certain outcomes. A meaningful example is what happened with the Occupy Wall Street (OWS) movement, a collective protest in September 2011 in New York City that targeted big financial organizations, advocated for more income equality in the US, and denounced the prominent role of money in politics. The protest attracted attention in the US and worldwide and became a movement (and a slogan) to reduce the influence of corporations in politics, achieve a more balanced distribution of income, and promote a reform of the banking/financial system to assist lower income citizens. Social media were used prominently to promote the initial protest in September: in fact, the activists intensely used Twitter to coordinate the protest and made the public aware of what was happening in South Manhattan, in the Wall Street district, where the protest originated. Surprisingly enough, the various hashtags used by the activists did not trend significantly, which means that, despite what was happening, Twitter (i.e., the algorithm—note that Twitter uses advanced ML systems to help surface relevant contents) did not seem to reflect the event's breadth and magnitude. According to a report from journalist Tarleton Gillespie,[2] some suggested that Twitter deliberately dropped the hashtag #occupywallstreet from its list of trending hashtags to prevent OWS from reaching a wider audience. The public initially reacted with anger toward Twitter and saw it as a political tool to control public opinion. Those who supported the protest were under the impression the social media company was aligned with the financial and political status quo which OWS was trying to challenge and, thus, that it conducted de facto censorship. However, Twitters' engineers denied any censorship but could not explain why OWS had not become a trending topic. Nor could they forensically trace back how all the data streams received by Twitter was processed by the social platform's algorithms. In other words, the algorithm was so complex and blackboxed that its output could not be explained by the very engineers who designed it and managers who implemented it to support social exchange and discourse.[3] The OWS example illustrates how relatively simple algorithms such as the ones that manage

[2] https://limn.it/articles/can-an-algorithm-be-wrong/.

[3] This example draws on Li et al. (2024).

Table 4.1 Examples of the Transparency vs. Blackboxed Continuum

Sociotechnical context	*Transparent*	*Nuanced/mix*	*Blackboxed*
Programmers/ designers and decision-makers	Uber performance management systems powered by ML	Hiring software (due to the lack of human supervision)	Twitter engineers not knowing why some hashtags didn't get traction during and in the aftermath of OWS
End users	FICO credit score to apply for a loan, people analytics in professional bureaucracies (high-tech, higher education, healthcare)	Uber drivers	Twitter algorithms, second degree nudging

news feeds in social media can become cryptic even to their designers. The more recent advances in ML systems, which learn from real world data and shape algorithms based on real-time input, make it even more difficult for designers to explain why AI systems come up with certain outputs.

To summarize, AI systems are powered by algorithms that can be more or less transparent. To whom? This question applies to programmers/ designers, decision-makers, and end users. Table 4.1 reports on these insights with examples.

In the reminder of this chapter, I focus on three contexts concerning algorithms where I collected primary data and that showcase the transparency-blackbox continuum. The first context is a healthcare organization in Massachusetts that reflects the difficulties (and successes) of implementing people analytics initiatives in professional bureaucracies. The second context concerns Uber drivers and their interactions with the "Uber app," an AI-based system that tells them what to do without providing explanation and feedback, which reflects the struggles of gig economy workers. The third context is about social media algorithms where I offer some explanations on why they can be discriminatory and why it is difficult to prevent the discrimination. For my arguments, I rely on a qualitative social media study that I conducted in 2015 with over 100 participants and an interview with a colleague who studies social media algorithms and fake news.

Workplace Surveillance in Professional Bureaucracies

Surveillance in the workplace is a phenomenon and trend that's been around for decades in various forms. The first notable example concerns Frederick Taylor and the scientific management principles in the early 1900s (Taylor, 1911). One century later, the people analytics literature has mainly focused on human resources (HR) as a tool to assess individuals using analytics capabilities. People analytics doesn't equate with surveillance. Performance management systems developed in bright light is quite normal for large companies where qualitative insights are not enough. Also, qualitative ways to assess employee performance might be highly subjective. For instance, management by objectives (MBO), which Peter Drucker (1954) first popularized, focuses on specific goals based on specific, quantitative metrics that employees need to achieve to receive rewards (or otherwise face punishment for not achieving them). At the time Drucker laid out MBO principles, many believed that this approach would fairly assess performance. Yet, one downside with MBO was that employees would pursue objectives at any cost even if they had to do so illegally or unethically. The key difference between performance management systems such as MBO and people analytics (or even the Tayloristic way to measure performance) is that the former is based on results (typical of professional bureaucracies) while the latter is typical of machine bureaucracies. In Chapter 2, I discuss machine bureaucracies when illustrating, using Amazon warehouses as an example, control focused on processes. Here I discuss professional bureaucracies, another concept that Henry Mintzberg developed in his seminal book called *Structure in Fives* (1983).

Professional bureaucracies include universities, high-tech companies, healthcare systems, and, more generally, organizations where people have sufficient qualifications (they are "professionals"). For instance, in universities, professors are evaluated for their research (in research institutions) or teaching (in teaching institutions) or for a combination of research and teaching portfolio (in "blended" institutions). Whether you are an adjunct instructor in a teaching university or a tenure faculty member in a research-intensive institution (called "R1 institutions"),[4] your manager (a department chair or a dean) will not look into your everyday activities: how you teach (also due to academic freedom, which the American Association of University Professors spells out in the US)[5] or how you research (how much time you write in a day or how many papers you submit in a month). In these types of (professional) bureaucracies, faculty members are assessed on outcomes. They are qualified enough to benefit from academic freedom, yet they need to achieve good student evaluations (for teaching faculty) or publish papers in recognized outlets (for research faculty). In high-tech companies, people work loosely because innovation processes need creativity, which is not tied to the number of hours/week someone spends in the office. Yet, if results are not achieved, punishments occur. In healthcare, doctors have a similar status to academic tenure, and their performance depends on their results according to specific outcome metrics.

For instance, in the US, the "value-based approach" has progressively replaced the "fee-for-service" approach since 2016. Healthcare.gov[6] defines fee for service as "a method in which doctors and other health care providers are paid for each service performed. Examples of services include tests and office visits." In contrast, a value-based system involves "linking provider payments to improved performance by health care providers." The value-based approach holds healthcare providers accountable for both the cost and quality of care they provide. "It attempts to reduce inappropriate care and to identify and reward the best-performing providers" (healthcare.gov). Because federal funds for hospitals match healthcare quality, it is of paramount importance that hospitals are in good standing with respect to taking care of their patients. For instance, not medicating a patient with diabetes with hypertension drugs (if needed) is a red flag (a hospital would lose points and funding). Doing otherwise could lead to long-term expenses for the whole healthcare system since a diabetic patient with hypertension is more likely than other patients to have a stroke or a heart attack, which both require long-term and expensive care.

[4] https://carnegieclassifications.acenet.edu.

[5] https://www.aaup.org.

[6] https://www.healthcare.gov/glossary/.

Deploying people analytics in professional bureaucracies is awkward by definition because people analytics control processes, not outputs (cf. Tayloristic approach). However, recent attempts to introduce this type of control made news. Microsoft Corp is among the few professional bureaucracies that tried to implement surveillance and process-based data collections. The company deployed a tool focused on line management that collects sensors data during meetings to assess productivity (using so-called productivity scores) based on various biometrics such as facial expression, tone of voice, gestures, and so on.[7] Yet, reactions to this tool included management's profound concerns because a "full-fledged workplace surveillance tool" might be counterproductive to organizational wellbeing.[8] In other contexts, data is collected through surveys (and on a voluntary basis) with potentially biased results (Leonardi & Contractor, 2018). In fact, even though surveillance is invasive, borders on being illegal, and might negatively affect workplace climate, it to capture behaviors more objectively than surveys.

One meaningful example of surveillance in professional bureaucracy comes from the intense use of Zoom.[9] Because of the increased use of remote work (in companies), online learning (in higher education), and healthcare during and after the COVID-19 pandemic, Zoom has become central to once face-to-face interactions. Yet, the analytics capabilities embedded in software such as Zoom are astonishing and, to some extent, creepy. Here I see two main issues. One relates to privacy and the other to discrimination. Note that privacy and discrimination are, in my opinion, the two key problems concerning IT ethics as I explain in a paper with fellow colleague M. Lynne Markus, Emerita Professor and former John W. Poduska, Sr. Endowed Chair at Bentley University (Marabelli & Markus, 2017).

One example of privacy issues concerns updated Zoom terms and conditions that the company announced in March 2023 that allow the company to collect data on Zoom interactions (recorded videos and

[7] https://www.geekwire.com/2020/microsoft-patents-technology-score-meetings-using-body-language-facial-expressions-data/.

[8] https://twitter.com/WolfieChristl/status/1331221942850949121?ref_src=twsrc%5Etfw%7Ctwcamp%5Etweetembed%7Ctwterm%5E1331221942850949121%7Ctwgr%5E%7Ctwcon%5Es1_.

[9] https://zoom.us.

chats) to train their AI system. As an article in *Wired* in August 2023 reported,[10] the new terms and conditions stated that customers:

> Consent to Zoom's access, use, collection, creation, modification, distribution, processing, sharing, maintenance, and storage of Service Generated Data [for purposes including] machine learning or artificial intelligence (including for training and tuning of algorithms and models).

After angry reactions to this news when it became public on social media outlets, Zoom backed off, and its chief product officer stated that: "We will not use audio, video, or chat customer content to train our artificial intelligence models without your consent."[11] While Zoom revised its terms and conditions, it remains unclear whether they can collect and use user data for training AI systems.[12] How can data harvested on Zoom hurt people? Who are the victims? In relation to companies, all employees can be tracked down with respect to their everyday work practices. Analytics capabilities available on Zoom can support the generation of evaluation metrics based on inaccurate data. For instance, the software can "tell" management who was more (and less) vocal in Zoom meetings. Zoom doesn't need to use voice-recognition systems to generate such reports. In fact, as many people know, in multi-way Zoom meetings, whoever talks can see the borders of their video become green. Should managers use this data to assess employee participation in meetings, they would, for example, potentially discriminate against women because they typically talk less in meetings (especially if the lead person is a man) and are more prone to turn their camera off (e.g., to avoid sexist treatment).[13] Discrimination stemming from Zoom analytics results from well-known gender unbalances. For instance, during the COVID-19 pandemic, helping children (or partially homeschooling them) tended to fall on women's shoulders. As a result, the pandemic exacerbated preexisting gender disparities in terms of job position and salaries: women became too busy to work online or even had to choose between either

[10] https://www.wired.com/story/fast-forward-generative-ai-companies-thirsty-for-your-data/.

[11] https://blog.zoom.us/zooms-term-service-ai/.

[12] https://news.ycombinator.com/item?id=37041632.

[13] https://www.nytimes.com/2020/04/14/us/zoom-meetings-gender.html.

homeschooling their children or performing salaried work (Malisch et al., 2020).

Other forms of discrimination arising from using algorithms to assess participation in Zoom meetings concern shy people or, more generally, people with social issues who might be less keen on speaking up (regardless of their gender). This doesn't mean that these people are less productive or pay less attention to Zoom meetings. Yet, metrics relying on who talks the most may penalize these individuals. Furthermore, relying on algorithms that make up performance based on active contributions to meetings doesn't do justice to the quality of contributions.

In higher education settings, Zoom's ability mining data can be good for attendance purposes and to penalize students with their camera off if it is mandatory to turn on the video during class (as it should be, I would add, unless students hand in a specific accommodation to the instructor). However, uncontrolled use of analytics to assess student performance might lead to issues (especially concerning gender discrimination) similar to the ones that I note above. Also, in attempting to respond to student demands to accurately assess class participation in Zoom classes at all times, I've heard colleagues considering using Zoom analytics to monitor breakout activity, such as when students work or perform tasks more generally in teams. To this end, I think that we need to consider whether this invasive approach is productive (i.e., a fair tool to assess participation) or whether it will simply inhibit students' spontaneous interactions because they know they are being observed at all time. Here too, one should not conflate interaction frequency recorded via Zoom analytics with contribution quality. The fact that quantitative and analytics-based performance management tools are flawed, because they don't capture the nuances of work practices is well-known. In season 8 episode 1 of the popular TV show "Suits,"[14] management at a law firm decide to use an algorithm to decide who to fire, in a downsizing effort. The algorithm looked at specific, outcome-focused metrics and led to the firing of someone who was not delivering as much as other associates, yet he was the "glue" of all associates, being a team player and one who, more than anyone else, was keen on helping colleagues. The episode ends with management understanding that dry metrics do not do justice to "soft"

[14] https://www.imdb.com/title/tt1632701/.

contributions, invisible to the algorithm they were using. Yet, companies still attempt to rely on these metrics, because convenient and time-saving.

In both organizational and higher education settings, Zoom analytics looks like a very transparent way to assess performance. This might lead to employees and students attempting to game the system. For instance, in companies, colleagues and teams might agree on running Zoom meetings solely to leave a recordable digital trace. As companies are supposed to disclose whether they control meeting activities, should employees be aware that their meeting activities are not monitored, they might decide to run "empty" meetings or meetings where a Zoom link is established but then the participants do other things. In higher education, students might purposely make noises with their microphone open specifically to have Zoom record their frequent interactions. These examples are obviously speculations and do not consider the ethical standard generally employees and students uphold. Yet, when designing systems that can automatically discipline, punish, or reward people, we need to also consider the possibility that people will identify workarounds or game the system.

To summarize, the widespread use of Zoom involves two important ethical concerns. The first is about Zoom using user data to train their own AI systems. The second is about users being surveilled, stripped of their personal data, and being assessed unfairly. While I discuss DEI issues concerning emerging technologies and people with disabilities more in depth in Chapter 6 on the metaverse, I note here that, for instance, people with disabilities might experience challenges when interacting with peers via Zoom.

While focusing on processes rather than outputs is an awkward practice in professional bureaucracies such as high-tech companies and universities, healthcare is one such setting where it is particularly challenging to evaluate employees (i.e., clinicians) for how they do things rather than what they achieve. One thing that makes it particularly difficult to, for example, assess doctors during their work activities is that hospital administration is often less knowledgeable about medical practices than doctors (it is not uncommon that, in large hospitals, executives running the hospital don't have a medical doctor degree).

In the period from 2016 to 2019, I conducted an ethnography study at Alpha (anonymized name), a healthcare network (a public hospital) in Massachusetts to examine its people analytics initiative. Despite (or because) the lack of funds—public healthcare in the US generally struggles financially because they receive very limited financial resources from the Federal Government yet deal patients who are, for the most, not insured and can't pay their medical bills—Alpha set up a people analytics initiative to improve care quality. In particular, the initiative focused on improving some specific metrics concerning treatment of people with diabetes and hypertension (among others). These metrics are acknowledged by the Federal Government, which releases funds to public hospitals based on their ability to achieve results. All this relates to the value-based system, which rewards carers that can medicate patients, avoid chronic diseases, and, thereby, save money in the long run. The people analytics initiative at Alpha also monitored process efficiency related to the "patient journey" from checking in (front desk) to checking out (and leaving the facility) for a routine doctor visit (annual physical exam). Management believed that if the patient experience were good, they would be more likely to follow up with prescribed medications (e.g., for diabetes and hypertension).

Among the multitude of insightful findings from this study (for details, see Marabelli et al. (2020), three are key and that relate to this book because they have important implications for how algorithmic-based systems are being managed in the workplace under a people analytics regime. First, professionals such as physicians are extremely uncomfortable when monitored for what they do (*how* they do things) rather than assessed for what they achieve. For instance, doctors at Alpha told me that they felt uncomfortable when told that their visits should be shorter. They had a point: as patients might need to spend more or less time with a doctor for a physical examination depending on their psychological needs. A Residence physician told me during an interview that some patients "need" to talk with a doctor for a while. This brings to the surface a substantial issue associated with quantifying, via people analytics, professionals' behavior. While I am not saying that it is OK to emotionally abuse Amazon drivers or warehouse workers by using Tayloristic practices, doing so in other contexts such as professional bureaucracies becomes even more problematic due to the high process variation in these contexts.

Second, I found that it is possible to leverage doctors' professionalism to improve their performance. At Alpha, management decided to print each doctor's performance results (related to time of visits and their ability to have patients take the drugs they needed) on the walls of the main hallway of the clinic where they worked (see Fig. 4.1).

Fig. 4.1 Performance assessment

A hospital manager told me that this strategy—exposing individual performance—worked well because it pushed doctors to be "more efficient" because they are inherently competitive. He told me that:

> They [doctors] all went to get an A, right, they went to college and they all got A's. Right. And they went to medical school and they all want to get A's. That's their personality. They wouldn't be doctors if they didn't want to get an A. So as soon as you put this up there and they go look at it and say how did he do or how did she do and they go oh, I can do better than that. You don't say anything but this taps in to that whole personality ... Doctors got to get an A, got to get an A, everyone of them. They got to get A's, that's their personality...

Competition among doctors, however, might lead to unintended consequences; for example, they could overlook patients in need to spend quality time with their physicians due to psychological issues. This was especially true at Alpha where most patients were homeless or otherwise displaced and at the margins. In other words, the hospital pursued an efficiency strategy, and one might argue that to some extent this happened, in part, at the expense of minorities (a population with a significant higher need for psychological and medical care). This raises an important tradeoff in the US public healthcare system. Alpha's administration pursued an efficiency strategy to obtain more federal funds in order to have more resources to take care of underserved communities in the long run. While at Alpha patients were never denied care, in these contexts, pursuing efficiencies to meet federal standards may lead administrators to provide a temporary disservice to such communities. This, in my opinion, is a case of potential discrimination, inequity, and exclusion from the top down, which pushes public healthcare organizations to defocus from short-term healthcare goals in order to secure long-term ones. In fact, the revenue-driven healthcare system, at the federal level, focuses on avoiding long-term medical expenses, which forces public hospitals to treat patients as non-human entities with the right to access limited clinical resources—the resources that would make them less likely to cost money to the government in the long run.

Third, during my three-year long ethnography study at Alpha, I was able to capture the longitudinal dimension of the people analytics initiative. While doctors initially expressed reluctance to being surveilled, one year into the initiative, they not only started to compete with each other to "look better" (they all wanted to "get an A") but also begun to proactively suggest to the administration how to measure their work practice in a way that could lead to efficiency but would also do justice to their responsibility as doctors (e.g., in reference to spending more time with patients with mental issues). This negotiation led to rigorous assessment of doctors' performance but in a way that departed from the typical coercive control of people analytics. A formative, bottom-up approach to analytics proved effective both for patients and the hospital's bottom line. It is nevertheless worth noting that these cases where the workforce engages in co-developing analytics systems is seldom pursued outside professional bureaucracies and, in my experience, are quite rare even in these types of organizations. Pratyusha Kalluri (2020) put forward this idea of bottom-up approach to build analytics systems where the underlying algorithms are built by the very people who are monitored. While my first reading of Pratyusha's *Nature* paper left me skeptical, deeper reflections made me think that the use of algorithms in professional bureaucracies can represent a starting point for bringing "end users" to the decision-making table. But implementing bottom-up people analytics initiatives is far from being a straightforward process.

One problem concerning the use of algorithms in organizations concerns information asymmetry and power distribution. Managers knows what they are measuring and how, while employees often do not. Managers primarily maintain this information asymmetry based on the fear that employees will game the system as they would understand what algorithms look into. For instance, in my study at Alpha, I learned that a way to track doctor visits was to record when the doctor (or a medical assistant) started documenting the visit on their computer using the electronic medical record (EMR) system EPIC[15] and when they eventually saved the record (start and end of the visit). In turn, it would be extremely easy for a doctor to delay logging into EPIC and perform a long physical examination unnoticed. Doctors initially saw this practice as a workaround to fix the unfair way the administration tried to purse efficiency.

However, at Alpha, frequent meetings between lead doctors and the administration led to a negotiation where clinicians began to participate in designing analytics systems to monitor their practices. Even if at first this could seem awkward or even as not making sense due to the obvious conflict of interest (how can surveillance be effective if the surveilled decide and know how they should be so), the collaboration between the two parties eventually worked well. Doctors agreed to be assessed by their performance with patients and even suggested new metrics that would make the monitoring more effective. Surveillance then turned into a self-assessment tool that benefitted both the administration and the doctors. How was it possible? First, in hospitals, doctors have "tenure" and cannot be fired because their visits take more time than they should. Thus, at Alpha, doctors felt comfortable "breaking the rules" when needed as they knew they would not face any consequences from doing so. But, at the same time, having automatic systems checking out their practices was an effective tool for doctors to self-assess their performance. And if a doctor

15 https://www.epic.com.

didn't perform well, the consequences were minimal. Also, it is worth noting that healthcare is a very particular setting. It is not only a professional bureaucracy (where people with "tenure" status cannot be easily laid off) but is also a context where workers see their appointment as not just as a job. Healthcare workers (doctors, nurses, technicians, and staff) are extremely committed to their job to help other people—particularly in the public US healthcare systems where doctors receive much less pay than in private hospitals.

Using people analytics in professional bureaucracies is challenging because professionals are not used to being surveilled. Zoom surveillance poses relevant privacy issues and DEI issues associated with people with disabilities. In higher education, students can be prone to the same DEI issues as company employees on Zoom in addition to the fact that quantitative evaluation is optional while in companies it is mandatory in college (grades). One example concerns disabilities: in a "Zoom world" where automatic assessment takes place, companies can waive people analytics initiatives to accommodate employee needs, while, in higher education, students with disabilities might be left behind. Healthcare is a unique type of professional bureaucracy where surveillance can negatively affect healthcare service delivery if used irresponsibly. With this, I don't mean to say that Amazon surveillance is an ethical way to manage the workplace. Amazon's oppressive surveillance harms its workers who face significant exploitation, with people often people at the margin with no means to defend themselves or who simply do not have an alternative to make ends meet.

Healthcare surveillance—at least in the case I describe above (Alpha)—affects workers minimally but can hold consequences on broader populations, such as the populations that public hospitals serve. The same populations that rely on public hospitals in the US are the people who work three jobs, take underpaid daily double shifts, and sleep in the very place where they operate all day long to make a near to survival income (another example includes Uber drivers[16] who don't benefit from insurance coverage from their employee (Uber) because they are contractors). For this reason, using people analytics in healthcare is a delicate matter. It can be very successfully if implemented in a bottom-up fashion as Alpha

[16] https://www.theguardian.com/technology/2019/may/07/the-uber-drivers-forced-to-sleep-in-parking-lots-to-make-A-decent-living.

did. But otherwise, the risk of clinicians operating with a focus on efficiency (and to meet goal-metrics set by the Federal Government) may be detrimental to healthcare quality, especially in contexts where the population involves, in large part, people at the margin. What is paradoxical is that, while rich, private hospitals can access a wide array of financial resources (along with federal support, they benefit from donors and often accept only insured patients; i.e., people who pay their bills), public hospitals often lack funding and need even more than private hospitals to meet certain federal-set performance thresholds to receive adequate funding. Furthermore, to meet these thresholds, they often need to implement metric and efficiency-based systems such as people analytics systems.

GIG Economy Surveillance

Gig economy workers have always been the target of mass surveillance for two main reasons. First, it is easy to employ monitoring techniques in settings such as online platforms or other settings where micromanagement happens through digital technologies and sensors (Amazon drivers, Uber drivers) because these technologies are designed to harvest data (e.g., data extractors). Second, gig economy workers often do not have an alternative. They don't choose to be exploited by companies who consider them "contractors" (e.g., Uber) even if they work more than 60 hours a week. The difference between contractors and employees is substantial in the US because the latter are eligible to various benefits, including healthcare coverage. Also, several employers do not allow unions. Indeed, we can see longstanding tension between Amazon employees who want to join a union and the company that refutes their wishes, such as by rejecting their requests often ironically by surveilling activists.)[17] Amazon recently started using gig economy workers for their delivery services.[18] Because gig economy workers often do not have an alternative, they are more likely to accept work conditions that involve mass surveillance. It is for these reasons that it is important to examine work conditions of gig economy workers and reflect on potential ethical and DEI issues.

[17] https://www.cnbc.com/2020/10/24/how-amazon-prevents-unions-by-surveilling-employee-activism.html.

[18] https://www.bloomberg.com/news/articles/2022-05-20/amazon-is-using-gig-economy-drivers-to-deliver-from-malls.

In the past few years, I conducted two qualitative studies on the Uber community with colleagues from various universities in the US (Mohlmann et al., 2023; Tarafdar et al., 2022). Both studies involved interviews with over 50 Uber drivers who discussed their struggles with using algorithmic-based systems that are supposed to help them perform their job but that actually controlled them. What I learned from my research on Uber drivers builds on Alex Rosenblat's (2018) popular book *Uberland: How Algorithms Are Rewriting the Rules of Work* in which she explains how Uber drivers are systematically exploited. While Rosenblat focuses mainly on drivers' work condition—something that is extremely relevant to capture the misery in which Uber puts its drivers—my research shifts the focus to psychological and relational aspects of the interactions between humans and the technology at hand. This focus gave me and my colleagues the opportunity to theorize around human–computer interactions (HCI) by shedding light on the bonding (or not) between drivers and their app. Uber drivers have a very conflicted relationship with the app they work with. They see it as a *digital boss* from whom they, however, do not receive feedback. Instead, they are evaluated, disciplined, and punished for reasons that the company never discloses. This goes back to the tradeoff between a blackboxed and a transparent system (cf. Marabelli et al., 2021). The fact that working for Uber means that no formative feedback is being released means that Uber drivers have no room to grow professionally. Yet, they do learn how to "look" better for the algorithms using workarounds. For instance, in the study with Monideepa Tarafdar and Xinru Page, we found that Uber drivers discuss their struggles (dealing with the Uber app) in forums and offer advice to each other on how to "perform well" often by gaming the algorithm.

All this is concerning and reflects the longstanding two-tier employment system that characterizes America. Privileged individuals (white collars) are full-time employees, receive health benefits (among other benefits), and receive formative feedback from their superiors—a condition to constantly improve, learn new things, and pursue new career paths. Other individuals do not constitute employees, cannot form unions, don't have health coverage (even if with the Affordable Care Act[19] things got better for the poor), and are not offered a career. These people are often

[19] https://www.hhs.gov/healthcare/about-the-aca/index.html.

part of segregated communities and are the victims of "surveillance capitalism" to quote Shoshana Zuboff (2019). One natural consequence of this segregation is collective action, which, in the case of Uber drivers, has led to countless contributions to online forums (just to give an idea, the only website Uberpeople.net hosts hundreds of forums where drivers share advice, exchange experience, etc.). *Algoactivism* (Kellogg et al., 2020) is a consolidated phenomenon developing in online communities and aimed at fighting controlling algorithms. *Algoactivism* reflects bottom-up initiatives of people who aim to have a say in how algorithms manage people in a way similar to what I found from studying Alpha (see above).

However, the bottom-up approach described in professional bureaucracies such as Alpha differs substantially from what happens with Uber (or other gig economy workers). At Alpha, long-term exposure to analytics surveillance led to the co-production of a system that pleased the administration and made the doctors willing (i.e., accepting) to be monitored in their everyday practices. In that context, the two parties worked toward the same goal: meeting metric thresholds set by the Federal Government in order to secure funds to improve healthcare service delivery and better assist underserved communities, people at the margin, and, more generally, jobless, uninsured minorities who can't afford medical bills. The gig economy is a machine bureaucracy on steroids. MTurk workers, labelers from the Global South, and gig economy workers in general are exploited and their "getting together" does not involve speaking with the very people who control them. In professional bureaucracies, power asymmetry can be addressed (and mitigated) via negotiations and the disclosure of how surveillance systems worked. In gig economy contexts, power asymmetry widens with workers trying to circumvent the "oppressors" with workarounds. America has a long history of oppressing workers and denying them basic rights, something that is carefully reported in Kim Phillips-Fein's book *Invisible Hands* (2009). Here we are witnessing patterns perpetuating historical injustices with a shift from the production line of General Motors, General Electric, and Du Pont to online platforms and gig economy workers. As I discuss more in Chapter 5, without clear regulations in the workplace and unions, high-tech companies can continue to exploit human capital and, in particular, people who cannot defend themselves but need to work to support their families.

Social Media Algorithms

The last context I want to examine in this chapter concerns social media settings and their underlying algorithmic-powered AI systems. In these contexts, algorithms lack transparency. As for why, social media are run by private companies that focus on gaining revenue and whose business revolves around ads (whose value depends on how much the companies know about their users). In addition, the field lacks significant regulations, and social media users use platforms mostly for fun (which allows companies to apply their surveillance rules and avoid disciplinary action, such as from laws that protect employers or contractors) and for free (so private companies can make their own rules). But there are caveats.

First, it is true that private companies do not equate with government services. However, because of the prominent role and influence of social media companies in our society, some argue that these platforms should be considered public utilities and should be disciplined as such.[20] A related argument concerns how some platforms operate in industries such as healthcare but are not upheld to healthcare industry rules and standards. For instance, Apple offers Apple Watch, a wearable device that the Food and Drug Administration (FDA)[21] approves, which means that its use is clinically recognized. Therefore, Apple must be HIPAA[22] compliant. But the myriad health apps used on smartphones are not covered by HIPAA, a loophole that app owners exploit to collect data in an unregulated way.

The underregulated nature of social media stems from two key issues. One issue concerns the slow pace at which the US Congress is regulating the industry. While I substantially expand on this issue in the next chapter (which specifically focuses on regulations), here it is enough to say that the idea that technologies evolve too fast, and regulators can't keep up with them no longer stands as we can see the difference between the US and Europe in regulating AI. To wit, it is possible to at least

[20] https://technosociology.org/?p=102.

[21] https://www.fda.gov.

[22] HIPAA (Health Insurance Portability and Accountability Act, https://www.hhs.gov/hipaa/) is a federal law that required the creation of national standards to protect sensitive patient health information from being disclosed without the patient's consent or knowledge.

attempt to prevent personal data misuse, discipline algorithmic transparency, and so on. The other issue concerns longstanding (and probably outdated) laws protecting social media companies (e.g., Sect. 230, part of the 1996 Communications Decency Act, which provides immunity for online platforms with respect to third-party user-generated content). Said in plain English, social media companies are not responsible (read: cannot be sued) for content that users post. While there are pros and cons concerning Sect. 230, it represents an example of the outdated laws regulating social media. For instance, Sect. 230 was written in an Internet era where algorithms didn't dominate the way users access platforms. In 1996, user-generated content related to discussion forums and chatrooms. Even Facebook, ten years later, started its platform by offering user-generated content in chronological order with users being exposed to others' contents in a somewhat "neutral" way, until soon after, when they implemented the EdgeRank algorithm.

Nowadays, users are profiled and offered targeted content (concerning other users' content, news, and ads) that often features discrimination. For instance, ads on Facebook often target users with job postings based on gender stereotypes. Indeed, a report by Global Witness reported on how Men overwhelmingly see mechanic positions, while women overwhelmingly se ads for teachers.[23] They took a closer look at France and the Netherlands and found highly discriminating ads as Fig. 4.2 showcases.

Second, it is true that social media companies provide access their platforms "for free."[24] Yet, trading data has its own price, and is appealing to data brokers, another largely unregulated market as I illustrate in Chapter 2. While money transactions are visible and traceable (or at least should be), such as when brokers sell records of plates crossing state borders to law enforcement to spot women trying to access abortion services,[25] the data that users provide to social media generate invisible transactions. This happens simply because social media users are not compensated for the personal data they give up when generating online

[23] https://www.globalwitness.org/en/campaigns/digital-threats/new-evidence-of-facebooks-sexist-algorithm/.

[24] Exceptions are Snapchat, that launched its fee-based Snapchat Plus service (2022); Twitter, that followed suit with "Twitter Blue" (2022); and Meta (Facebook and Instagram, in 2023), that test-marketed a subscription bundle for their services.

[25] https://www.wired.com/story/license-plate-reader-alpr-surveillance-abortion/.

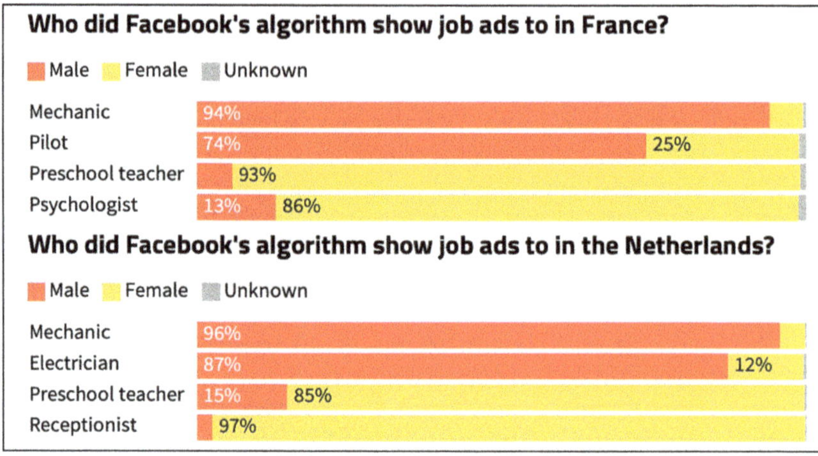

Fig. 4.2 Ads discrimination in France and the Netherlands (https://www.Globalwitness.org)

content. In addition, user data is often employed by social media companies in ways that deviate from the purpose for which the data was collected in the first place. One example is social media colleting gender and race data at the registration point for statistical purposes and then using the same data to deliver targeted job ads based on gender or race.

The race to collect user data via social media leads to workarounds that showcase people's willingness to react to algorithmic control. For instance, in 2015, in collaboration with Jerry Kane, the C. Herman and Mary Virginia Terry Chair of Business Administration and Professor in the Department of Management Information Systems at the University of Georgia, and Xinru Page, Associate Professor of Computer Science at Brigham Young University, I conducted a broad study on online behavior of nearly 100 social media users. The study led to several papers (Marabelli et al., 2016; Page & Marabelli, 2017; Page et al., 2016). As it relates to this chapter, the study is interesting because it reports people's awareness of the "power of the algorithms" in shaping their online experience even if it dates back to almost a decade ago. For instance, we asked the participants how they felt when skimming posts from Facebook's news feeds. In response, one participant said:

> If there were anything to make me not want to use Facebook anymore it would be that, it would be how annoying that is and how limiting that is. Because obviously you know you have to be pretty blind to not to see that that's what's happening. But it makes you feel like you know trapped, like they're limiting what it is that you could possibly stumble upon just because there's an algorithm out there that says that you'd maybe be less likely to click on it. So for whatever reason it's not presented to you. And that's frustrating and it's annoying.

Another study participant explained an attempt to contrast the power of the algorithm with practices aimed at confusing it:

> So, basically with Facebook I have all the friends there. They post their pictures. I know what they do. And I usually start liking things. But because I know that people follow Facebook and then companies collect information on Facebook, what I do is I usually, if I like pictures, let's just say I like ten pictures, so five of them I say that I like, which is true, and then the other five I say that I don't like, which is not the truth, because I don't want the companies to collect my profile on Facebook.

These two examples reflect the deep awareness of how social media manipulates posts ultimately to increase engagement, i.e., leading people to spend the most time on social media, a proxy for pricing ads along with the ability to demonstrate profiling capabilities. Facebook started manipulating its News Feeds feature in 2009, but it began building algorithms focusing on providing quality content to users since 2013 based on Facebook's Chief Operations Officer (COO) Sheryl Sandberg's strategy to provide content that would keep people on the platform for as long as possible. Today, we have evidence of the discriminating nature of the Facebook algorithm (e.g., in the way it offers different ads to men and women). One might wonder why we keep being so accepting of social media managing what we see and from whom (now including ads, news outlets, and even fake news). But probably an even more meaningful question concerns why, despite all reported evidence of algorithms discriminating across populations, social media companies perpetuate these practices.

A first explanation, which I consider true but also relatively naïve, is that social media companies are poorly regulated as I have already said and as I discuss more in the next chapter, which is all about algorithms and regulations/regulators. But here I will elaborate on another point that partially explains why social media companies continue to perpetuate discriminatory practices. I discussed this topic with Dr. Noah Giansiracusa, Associate professor of Mathematics and Data Science at Bentley University who studies algorithms and AI in social media settings, with a specific focus on social media, such as Facebook, that make use of recommendation algorithms. The overarching question we debated is: Do social media have to discriminate to be effective with their targeted ads? This question is not a trivial one, and led to a lively discussion, which I synthesize below.

The obvious premise is that it is possible to correct discriminatory algorithms that use historical data, and the example I focus on here concerns ads (e.g., Facebook ads as I discuss above). Generally speaking, social media algorithms are designed to engage users with content that they will like. A primary (and somewhat negative) consequence of this is the so-called filter bubble by which people receive mostly content they like and "approve." The algorithm filters out contents that might be not in line with one's thoughts and beliefs. In a famous Ted Talk in 2011, Eli Pariser was one of the first people who made this concept accessible to the general public.[26] According to filter bubble principles, social media algorithms do not just offer contents people are interested in, e.g., someone who likes dogs will see posts concerning dogs or animals in general. The bubble also concerns ideologies. Liberal people will see related posts (from news outlets and people who share their ideas). The same happens to conservatives. The result is that posts and news trends that are not in line with one's ideology are simply filtered out by algorithms. For a liberal, posts from a conservative "Facebook friend" become invisible. The main consequence of the filter bubble, as I explain in a paper from a few years ago (Newell & Marabelli, 2015), is that people who spend a lot of time on social media become less exposed to issues they don't agree with. They can already choose what news outlets to watch, what newspapers to buy, and, with social media algorithms constantly pleasing them with news, the filter bubble risks leading to settings with individuals becoming less tolerant to different opinions, ideas, and so on. In the long term, this can lead to polarized views and intolerance.

But how do social media companies know that we are liberal, conservative, or otherwise? They know what we like because the underlying

[26] https://www.ted.com/talks/eli_pariser_beware_online_filter_bubbles.

algorithms mine our online behaviors and profile our preferences—the basis of ads. Sticking to Facebook as an example, once one *likes* articles from liberal outlets, liberal sources, and even comments made by people who have already been "tagged' as liberal by the algorithm, the social media user becomes the target of related content, a clear manifestation of *classification*. In ads, this translates into the algorithm knowing what our interests are and inferring what we might want to buy because, e.g., someone who often visits Facebook pages related to dogs and likes others' photos of dogs will be offered ads concerning dog food, dog beds, groomers in their areas, and so on. So far so good. Most people are probably happy with a "friendly" social media that offers products and services in line with their interests—what would be the point of seeing things we are not interested in or being offered products and services that we don't want? While too much ads targeting might lead to nudging and leading people to buy things they don't need, overall ads targeting is something that I don't see as too negative. When asked if he'd rather use a social media without algorithms offering contents close to his interests, Dr. Giansiracusa agreed and said that: "I, as a person, I'd rather an entertaining TikTok than a boring one" (TikTok, in Europe is currently offered with a "no-algorithm" option, which can be chosen by users to prevent addiction).

However, problems emerge when *recommendation algorithms*—algorithms that process filtering providing items that are most pertinent to a particular user or group of users—are not just about dogs and dog food, but also concern job ads, opportunities to purchase a house (instead of renting an apartment), and so on. Let's now shift from jobs to housing, where Facebook's recommendation algorithms are well known for operating in a racial discriminatory manner.[27] To wit, the algorithm offered ads concerning homes for sale mostly to Whites, while Blacks where showed ads to rent apartments. This happens because Facebook's algorithm inferred that Black Americans may have less financial resources than Whites and, thereby, would be more inclined to rent rather than buy real estate. This practice leads to a situation where Black Americans have fewer opportunities to purchase a house not just because of their financial condition but also because they come across fewer opportunities to buy a real estate.

[27] https://www.propublica.org/article/facebook-advertising-discrimination-housing-race-sex-national-origin.

This echoes mid-1950s segregation practices (now illegal) where blockbusting (real estate agents persuading Whites to sell their properties at below market price in neighborhoods that will likely be populated by Black Americans—often not true—with the aim to then sell the same houses to Blacks at inflated prices) could operate because the FHA (Federal Housing Authority) made sure that Black Americans had few alternatives where to purchase homes at fair market value (Rothstein, 2017). Eighty years after blockbusting and FHA's segregation practices, we are witnessing algorithms doing more or less the same thing by preventing Black Americans from accessing real estate offerings. Instead, they are shown options to rent. Public data from the US Census Bureau[28] indicate that the median income in 2022 for White families was around US$75,000 in 2022 and around US$53,000 for Black families $53,000. This is already a huge difference. But median White household wealth (assets minus liabilities) was about US$187,000, while the corresponding median for Black household wealth was US$14,100 (i.e., less than one tenth of that of Whites). Because equity that families have in real estate is the main source of wealth for middle-class Americans, Black Americans remain disadvantaged as their parents and grandparents were not able to leave them properties. They were not able to benefit from the economic booms of the 1950s and 1960s, which instead allowed Whites to accumulate equity.

Going back to my initial question of why social media continue to use algorithms that perpetuate discriminations, the answer is that, to correct these algorithms, we need to deal with a tradeoff between discrimination and privacy. Technically, the reason why algorithms recommend houses for sale prevalently to Whites rests on the fact that most Whites click on links, ads, and *like* content that make the algorithm infer that they have more financial resources than other social media users (Black Americans, who have different social media behaviors. Is that true? Maybe. On average, Black Americans make less money than Whites in the US (see data from Chapter 1), but this is due to historical reasons associated with slavery in the 1800s, segregation, lack of applications of the Fourteenth Amend in most of the 1900s, and structural racism nowadays, of which recommendation algorithms are a good example. How to correct

[28] https://www.census.gov.

algorithms? Dr. Giansiracusa gave me an insight that allows me to infer potential solutions:

> The algorithm says, ah, based on everything that Marco has done, like his friends and the college he's gone to and all these kinds of information, I think Marco is very likely to click on this ad, so I'm going to send this to him. But it doesn't have to even know your race or anything. So that's the challenge, is that it's not using your race, it's using everything it knows about you. But guess what, Marco? What college you went to, the fact that you're a professor, what zip code you live in, who you're friends with, all of these things correlate with a lot of gender, race, educational background, a lot of these things, person, minority person, not a disabled person, and that data is what's used. So it's hard for the algorithm to know when it's discriminating. It just sees user A and user B and says, I think user A is more likely to click this rather than user B. It doesn't know if it's doing that because of bias or because it's clever and it figured out what user A is really interested in. Maybe user A is interested in cats and User B is interested in dogs, and this is an ad for cat food.

To correct an algorithm that makes recommendations based on online behavior, we need to equip it with information that goes behind behaviors. We need to feed algorithms with personal, sensitive data such as race, gender, disabilities, income, and so on. In this way, it can offer "neutral" ads at least with respect to critical topics such as jobs, real estate, education, and healthcare. First, it is important to single out "critical" categories where recommendations can lead to discrimination. For instance, healthcare is a domain prone to discrimination. Algorithms should be trained to offer cutting-edge cures and expensive drugs to everyone. Ads on food for cats or dogs could still rely on user online behaviors.

Second, in critical categories, we would need to collect sensitive data. Here is where the problem lays. If one wants to really make algorithms neutral on social media, they need to know sensitive data about users to correct algorithms that would follow trends driven by online behaviors rather than social justice. For instance, Black Americans should be offered the same real estate ads as Whites but, for an algorithm to do so, it needs to know that someone is a Black American.

This practice, besides exposing a tradeoff between discrimination and privacy, has two caveats. First, several Black Americans might specifically look to rent (because they can't afford to buy). Giving them advice that is not consistent with their buying power may become pointless—why would one waste time reading ads for million dollars houses if they can't afford them? To this end, while ads concerning jobs are without question unfair when they use data on users as proxy for gender and racial discrimination, real estate ads and the extent to which algorithms offer them are a more ambiguous matter. Second, it is problematic to collect race data, especially among minorities, because they know that they are/might be discriminated against. It is problematic to justify data collection on race for the sake of avoiding discriminations because harvesting data on race is discriminating in its own right.

Fair algorithms should be neutral at best, which means they should offer ads to social media users regardless of their race and gender (or regardless of other potentially discriminating issues, such as income, education, disabilities). Yet, this line of thought will compromise social media companies' ability to target ads, which is the core business (and main revenue stream) of these companies. Some targeting ads are detrimental to fairness. For instance, well-paid job ads showed to only men perpetuates longstanding male privileges. Other (targeted) ads are more appropriate. For instance, showing ads of jobs behind a desk to people with limited mobility are appropriate and would be near to insulting offering someone with walking disabilities ads of a treadmill for sale. This is why it is important to generate categories or classifications of ads and populations.

Classifications can be detrimental to fairness because they place people in categories and doing so makes it easier to discriminate (see Chapter 1 and the pitfalls of classifying, especially with digital systems). However, in some cases, classifications might be helpful to address important issues, such as social justice. Still referring to real estate and the discriminatory Facebook ads, they originate from past classifications concerning race and that led to slavery and de jure discrimination in the beginning and then, after the 1968 Fair Housing Act, de facto discrimination and segregation (specifically disadvantaging Black communities). So, nowadays, we need classifications to attempt to recover from a status quo that reflects past injustices with the result that the US is among the countries with the biggest difference between the rich and the poor. Classifying ads with respect to the extent to which they can negatively affect existing disparities could be a way to partially make up for the past. The caveat is that, to do so, social media companies needs to be allowed to collect even more data than they already do; otherwise, "[social media algorithms would] just see user A and user B and say I think user A is more likely to click this rather than user B" to quote Dr. Giansiracusa.

Conclusions

Algorithms and AI systems can be more or less transparent for designers, decision-makers, and end users. This applies to the workplace, various collaborations (e.g., gig economy workers), and social media. While it is not true that algorithms are always transparent for people in power (the designers and the decision-makers) as with Twitter (now X) and the OWS movement in the past, generally power asymmetries surface with respect to workers, contractors, and social media users being the victims of algorithmic opacity. Initiatives materialize into individual workarounds to circumvent the algorithms and into collective initiatives (Uber forums, employees getting together to protest against their bosses). Algorithmic opacity has nearly always negative consequences. In work settings, it generates unhealthy work environment (e.g., Amazon warehouses) and frustration (e.g., professional bureaucracies). In other types of collaborations such as in the gig economy, the result is *algoactivism*, and here too frustration is a shared sentiment. In social media settings things are a bit different as people join social media not to make money (for the most part) but for fun and to expand their social network. In this setting, algorithmic opacity is less perceived but has deeply worrying consequences that lead to discrimination. In the next chapter, I put forward potential solutions to all this by starting from the idea that government intervention can mitigate the negative effects of the widespread use of algorithms in work and social contexts.

References

Drucker, P. (1954). *The Practice of Management*. Harper Business.

Kalluri, P. (2020). Don't Ask If Artificial Intelligence Is Good or Fair, Ask How It Shifts Power. *Nature*, 583: 7815, p. 169.

Kellogg, K., Valentine, M., & Christin, A. (2020). Algorithms at Work: The New Contested Terrain of Control. *Academy of Management Annals, 14*(1), 366–410.

Leonardi, P., & Contractor, N. (2018). Better People Analytics. *Harvard Business Review* November-December issue. https://hbr.org/2018/11/better-people-analytics

Li, J., Litvinova, Y., Marabelli, M., & Newell, S. (2024). Ethical Implications of AI Use in Practice for Decision-Making, Research Handbook on AI and Decision Making in Organizations. In I. Constantiou, Joshi, M. P., Stelmaszak, M. (Eds.), *Research Handbook on AI and Decision Making in Organizations,* Edward Elgar Publishing Ltd.

Malisch, J. L., Harris, B. N., Sherrer, S. M., Lewis, K. A., Shepherd, S. L., McCarthy, P. C., Spott, J. L., Karam, E. P., Moustaid-Moussa, N., & Calarco, J. M. (2020). Opinion: In the Wake of Covid-19, Academia Needs New Solutions to Ensure Gender Equity. *Proceedings of the National Academy of Sciences, 117*(27), 15378–15381.

Marabelli, M., & Markus, M. L. (2017). Researching Big Data Research: Ethical Implications for Is Scholars. *Americas Conference of Information Systems (AMCIS).*

Marabelli, M., Newell, S., & Handunge, V. (2021). The Lifecycle of Algorithmic Decision-Making Systems: Organizational Choices and Ethical Challenges. *Journal of Strategic Information Systems, 30*, 1–15.

Marabelli, M., Page, X., & Kane, G. C. (2016). Understanding Social Media Use: How Far Will College Student Sampling Take Us? In *Academy of Management Meeting*.

Marabelli, M., Vaast, E., & Carlile, P. R. (2020). Making Lemonade: Dealing with Analytics Surveillance in the Workplace. In *Academy of Management Annual Meeting*. https://papers.ssrn.com/sol3/papers.cfm?abstract_id=3572363

Mintzberg, H. (1983). *Structure in Fives: Designing Effective Organizations*. Prentice-Hall Inc.

Mohlmann, M., De Lima, A., Salge, C., & Marabelli, M. (2023). Algorithm Sensemaking: How Platform Workers Make Sense of Algorithmic Management. *Journal of the Association for Information Systems, 24*(1), 35–64.

Newell, S., & Marabelli, M. (2015). Strategic Opportunities (and Challenges) of Algorithmic Decision-Making: A Call for Action on the Long-Term Societal Effects of 'Datification.' *The Journal of Strategic Information Systems, 24*(1), 3–14.

Noble, S. U. (2018). *Algorithms of Oppression, in Algorithms of Oppression*. New York University Press.

Page, X., & Marabelli, M. (2017). Changes in Social Media Behaviors During Life Periods of Uncertainty. *The 11th International AAAI Conference on Web and Social Media (ICWSM-17).*

Page, X., Marabelli, M., & Kane, G. C. (2016). College Student Sampling: Can It Represent Social Media Use by the Work Force? *22nd Americas Conference on Information Systems (AMCIS).*

Phillips-Fein, K. (2009). *Invisible Hands: The Businessmen's Crusade against the New Deal*. W. W. Norton.

Rosenblat, A. (2018). *Uberland: How Algorithms Are Rewriting the Rules of Work*. University of California Press.

Rothstein, R. (2017). *The Color of Law: A Forgotten History of How Our Government Segregated America*. Liveright Publishing Corp.

Tarafdar, M., Page, X., & Marabelli, M. (2022). Algorithms as Co-Workers: Human Algorithm Role Interactions in Algorithmic Work. *Information Systems Journal, 33*(2), 232–267.

Taylor, F. W. (1911). *The Principles of Scientific Management*. Harper & Brothers Publishers.

Zuboff, S. (2019). *The Age of Surveillance Capitalism: The Fight for a Human Future at the New Frontier of Powe*. PublicAffairs.

CHAPTER 5

Institutional Inertia and Corporate Sovereignty

In a recent paper I wrote for the *Journal of Strategic Information Systems* (for which I currently serve as a senior editor for full disclosure), I broadly discussed, among other issues that emerging technologies raise, the role of governments in regulating AI (Marabelli & Newell, 2023). At that time, I felt sympathy for lawmakers struggling to catch up with the most recent technologies in order to regulate them for two main reasons. First, lawmakers often lack sufficient knowledge about newly designed, developed, and deployed technologies. Second, governments used to lead in the discovery of new technologies and innovations. I still stand with the second claim that I made in the article for which I report a relevant extract below:

> While, in the past, governments used to lead scientific research on technologies (cf. in the US Arpanet and the release of internet to companies and the general public), nowadays the opposite happens. It is large companies, with more resources (and data), that lead technological progress. Examples span from the new race to space led by SpaceX, Blue Origin and Virgin Galactic (Pekkanen, 2019) to the development of algorithmic-based AI systems (Zegart, 2022) and the advances in information security and data protection. For instance, in the aftermath of the 2015 San Bernardino (California) domestic terrorism shooting, the US government was not able to hack the iPhone of one of the shooters' and hired a private company

M. Marabelli, *AI, Ethics, and Discrimination in Business*, Palgrave Studies in Equity, Diversity, Inclusion, and Indigenization in Business,
https://doi.org/10.1007/978-3-031-53919-0_5

> to do this (Levy, 2016). What is more, the leading role of private companies is not mitigated by regulations protecting individuals from disclosing uncomfortable research findings, and this concerns whistleblowers.[1]

However, I recently began to question my first claim. It is true that lawmakers are challenged when asked to generate laws on matters they know little about (e.g., the Internet represents a pertinent example when it first appeared). However, this "lack of knowledge" really became evident with social media. When Sheryl Sandberg's crew started to develop algorithms to manipulate Facebook's news feeds, in 2009, lawmakers worldwide weren't able to foresee the massive consequences of these changes for social media users. In fact, it took almost 15 years for the US Congress to call out Meta's algorithms on Instagram, one of the company's leading platforms.[2] But, nowadays, the public at large has learned how social media companies manipulate user feeds in ways that can jeopardize people's wellbeing. Indeed, as an example, a recent *Wall Street Journal* investigation shed light on internal research results at Instagram and showed that its algorithm led teenagers with eating disorder to visit Instagram pages that suggested that they keep losing weight. Furthermore, since 2020, Facebook (another Meta company) conducted internal research concerning how photo-sharing practices affected teenagers. As reported by the same article in the *Wall Street Journal,* "the company's researchers found that Instagram is harmful for a sizable percentage of them, most notably teenage girls." It was only because Frances Haugen (a whistleblower), after leaving the company, informed the press about Meta's internal research, which would have never become public otherwise—something that should make us reflect on the relevance of whistleblowers and the lack of protection they have.

All this is to say that, in the last decade, public knowledge about the consequences of algorithmic-led manipulation of user-generated contents on online platforms has become evident to most people. Main consequences involve harming individuals (as per the case in the *Wall Street Journal* report above). Even if it is true that lawmakers might not know all the details of biases and discrimination issues associated with emerging

[1] Marabelli and Newell (2023, p. 8).

[2] https://www.reuters.com/technology/us-senators-write-zuckerberg-after-report-instagram-algorithm-2023-06-30/.

technologies (in particular, at the time of this writing, concerning generative AI), we can't ignore the different approach taken by the US and Europe with respect to invasive and pervasive algorithms, which are often tied to AI. For instance, as I explain later in this chapter, while the EU has taken the side of citizens with the AI Act, the US is still struggling with generating mandatory regulations aimed at protecting people from threats posed by Silicon Valley companies that, while sanctioned and fined in Europe, operate freely in their home country at the expense of people's privacy and rights. But before delving into the nuances in the laws and regulations of the three major AI leaders worldwide (i.e., the US, Europe and China) in which AI regulations concerning citizen privacy have proliferated in the last two years (at least to discipline private companies because governments can still access user data), I provide some background on why the US seems to be so behind in dealing with the unwanted outcomes of AI, which, in my opinion, rests on historical reasons dating over a century ago if not more.

Historical Perspective in the US: Laissez faire vs. the New Deal

Today, the US represents the quintessential capitalist economy. According to Wikipedia:

> Capitalism is an economic system based on the private ownership of the means of production and their operation for profit. Central characteristics of capitalism include capital accumulation, competitive markets, price systems, private property, property rights recognition, voluntary exchange, and wage labor.

While I am not against capitalism in principle, I embrace Shoshana Zuboff's (2019) idea that, in the age of ubiquitous computing, forms of surveillance capitalism are likely to develop that reflect the widespread collection and commodification of personal data by corporations. Yet, in America, this way to manage personal data, which affects, for the most, marginalized people generating inequalities and social injustice stems from the country's history over the last century. Silicon Valley companies are now able to predate people's data in a way that is backed up by a system where laws regulating its collection and use are lacking. Data collectors have a role in our society only because they are able to harvest data on million Internet users or otherwise people whose social actions become digitalized (e.g., because they withdraw money from an ATM, because a camera catches them on the street, or because they have a smartwatch or smartphone). While data harvesting happens worldwide, the extent to which it is regulated (and the extent to which these regulations may prevent tech companies targeting people at the margin) is hugely country specific. The US, where AI serves businesses at the expense of citizens, is a good example of how capitalism translates into unregulated data exploitation. Among citizens, some can buy their privacy and can avoid being discriminated against (see my example of onboard devices and insurance companies from Chapter 2 and—a previous paper I coauthored [Marabelli et al., 2017]). However, many cannot buy their privacy and avoid discrimination due to their socioeconomic status, which, sadly, is often a proxy for gender and race discrimination. For instance (US data), in 2022, women earned 17% less than men. Black and Hispanic women in rural areas made only 56% of white non-Hispanic man in the same geographical areas.[3] Examples of this kind proliferate in the US.

To understand how disparities have grown in American society over time (and how they affect the use of AI and its lack of regulations today), it is important to take a historical perspective. One hundred and fifty years after the Declaration of Independence (July 4, 1776), the first modern democracy recognized that freedom (in this specific context, free markets) would transcend social justice. During the presidency of Franklin D. Roosevelt—in particular in the 1933–1939 timeframe—the New Deal was conceived and deployed. The New Deal comprised several programs, public initiatives, financial reforms, and regulations aimed at supporting farmers, the elderly, unemployed people, and youth. Reforms under the New Deal are commonly referred as the "three Rs" (**relief** for the unemployed and for the poor, **recovery** of the economy, and **reform** of the banking system) and were designed to prevent another depression such as the one that occurred in 1929. Despite all the caveats involving the typical American culture devoted to meritocracy and the principle that whoever is poor deserved to be—something that poorly accounted for the history of slavery and contemporary discriminations (cf. structural racism), the New Deal was a step toward social justice. The New Deal, as it was initially conceived, represented quintessential democrat principles

[3] https://www.forbes.com/advisor/business/gender-pay-gap-statistics.

about, among other things, strong government intervention in social life. In the 1930s, predominantly Republicans and Conservatives opposed the New Deal. Indeed, in contrast to the New Deal's pillars, Republicans and Conservatives wanted to lower taxes to defund the government, deregulate industries to prevent Unions from forming and, thus, to strip basic rights from workers (e.g., strikes) and to keep wages as low as possible to favor corporations, and, overall, support the idea that individuals become poor not for historical reasons but due to their laziness and wish to rely on welfare (Phillips-Fein, 2009). Matthew Desmond (2023) unpacks the idea of the poor as "guilty" well and focuses on why the richest country on Earth also has widest gap between the poor and the rich.

Scholars such as Milton Friedman (Nobel Laureate in 1976) influenced (and strongly supported) President Reagan's laissez faire [what] in the 1980s. Friedman was a strong advocate of low taxation, privatization, and deregulation, which is clearly at odds with Keynesian's principles that rely on the relevance of the state to pursue economic growth. In my view, the fact that America's modern history (say, for the past century) has swung between laissez faire principles and government intervention has generated an array of inconsistent laws and regulations along with contradictory policies. One example is Section 230, a section of Title 47 of the US code that was enacted as part of the 1996 Communications Decency Act of 1996 (Title V of the Telecommunications Act of 1996), which provides immunity for online computer services with respect to user-generated content.[4] Because of the intensive use of algorithms on social media platforms, Section 320 was questioned with respect to whether algorithmic-led recommendations should be protected by the law; Section 230 shields social media companies, which means they can't be called out for whatever people post (e.g., on YouTube). But what if inappropriate content posted by users is offered to someone who becomes influenced by this content via an algorithm? This is precisely the case of Gonzalez vs. Google,[5] a US Supreme Court case that has dealt with the question of whether or not recommender systems are covered by liability exemptions under Section 230. In 2015, 130 people were killed in a coordinate terrorism attack happened in Paris. The Islamic State took responsibility for such attack. Nohemi Gonzalez, an US citizen, was killed

[4] https://uscode.house.gov/view.xhtml?req=(title:47%20section:230%20edition:prelim).

[5] https://www.supremecourt.gov/opinions/22pdf/21-1333_6j7a.pdf.

in the attack. She was in Paris under an exchange program with California State University at Long Beach. Nohemi's family sued Google, YouTube's parent company, based on the argument that it was through YouTube's recommendation algorithm that YouTube led users toward recruitment videos for the Islamic State and that were partially responsible for their daughter's death. Eventually, the US Supreme Court ruled in favor of Google that defended itself by relying on Section 230, which protects Internet companies from damages created by user-generated content.

While both sides of the aisle have proposed changes to Section 230 over the past decade, nothing has happened precisely due to the huge polarization in US politics, something that many attribute to the "Trump era" and social media. I respectfully disagree as the swing between government intervention and laissez faire principles has been a characteristic of US politics for over a century. What is even more worrisome is that US Congress' incapacity to put forward clear laws and regulations (e.g., a revision of Section 230 but also the various attempts to regulate AI and algorithms) has an unpleasant consequence: we are going toward a trend by which it will be the US Supreme Court ruling on relevant matters concerning emerging technologies such as generative A, and its ethical and DEI implications. I expand on this issue later in the chapter when discussing the dysfunctional government system in the US.

Going back to the issue at hand—(free) market vs. government—I disagree with Friedman's assumption that the market can serve customers better than the state, especially in the current AI-dominated era, where user data is up for grab and the US deregulation with respect to data brokers allows government agencies to police harmless citizens, such as women seeking abortions in other states. Allowing brokers to sell data concerning car plates crossing state borders to law enforcement makes it easy to target people at the margin who seek an abortion to save their own life (Berg & Woods, 2023). The paradox here is that government's regulations are broad when targeting minorities (overturning Roe) yet very relaxed on matters related to protecting the victims of such overturning since they scarcely discipline brokers or restrict their efforts to collect and sell data in massive amounts. My investigation into US privacy laws that potentially limit data brokers from helping law enforcement prosecuting minorities—e.g., women seeking for an abortion out of state by driving a car because rich people, often not fitting the profile of a struggling woman at the margin, can take a plane—led nowhere; in other words, I found no laws protecting the privacy of people's "plates." My research was confirmed by Dr. Khiara Bridges, a Professor of law at UC Berkeley School of Law, who told me that:

> Our privacy is contingent on where we are. The Court has held that when you're in a public place, you likely don't have a reasonable expectation of privacy, and driving on a highway is paradigmatic public place. So, I can't think of why it would be illegal for anyone to make a list of all the plates that have been traveling on the Mass Pike and send it to someone else.

Murray Weidenbaum (among others, including e.g., Ayn Rand) became famous for his studies on the economic impact of government regulations. In his 1970s writings, he claimed that the US was undergoing what he called the "second managerial revolution" and argued that the US Government was expanding its role and power over the economy through regulations that would prevent the free actions of private companies. These thoughts were used in the early 1980s by President Reagan to persuade Americans that government intervention equates to socialism. Of note, at that time, the Cold War was at its peak, and it was easy to leverage Americans' fear about socialism spreading in the West. The US Government's refusal or inability to make changes laws such as Section 230 in the Communications Decency Act of 1996 or regulate some things (abortion) but not others (data brokers) reflects old practices concerning institutions pretending to promote deregulation while actually regulating for their own (partisan) interests.

As often implied in this book, history explains current facts. Think about schooling segregation in the mid-1900, which has continued almost to the present day. Until the 1950s, redlining practices were embraced by most (both private- and government-backed) financial systems. For instance, the Federal Housing Administration (FHA), founded in 1934 by President Franklin D. Roosevelt (as part of the New Deal), supported first home buyers for most of the twentieth century by securing otherwise too risky mortgages with banks that could offer loans to medium-to-low-income buyers.[6] With this insurance arrangement, banks assumed a diminished level of risk, which allowed them to offer a larger number of mortgages and with down payment as low as 3.5% of the purchase price (initially, the down payment had to be 10% against the purchase). First, this practice of using federal money to help

[6] https://www.hud.gov/buying/loans

(some) people become homeowners automatically discriminates against minorities of poor Americans who cannot afford to buy a house because the (federal) financial resources devoted to back mortgages of people who can afford to buy a house are taken out from initiatives aimed at building or paying for public housing for homeless people.

While the idea of creating an agency that could help the middle-class purchasing a first home is not bad on paper, the problem is that FHA's redlining practices prevented most Black Americans from purchasing a home, particularly in the period between 1945 and 1959. During that period, Black Americans received less than 2% of all federal insured home loans (Hanchett, 2000). What is most ironic is that, to prevent racial integration, redlining did not just penalize Black Americans but also White Americans wanting to purchase houses in Black neighborhoods. As an example—well documented by Rothstein (2017)—in 1954, in East Palo Alto (CA) the FDA not only refused to insure mortgages for Black Americans in specific White neighborhoods but it did not insure mortgages for Whites in neighborhoods where Black Americans were present, which de jure promoted segregation in small towns. In addition, several despicable practices took place, which local housing authorities (city councils and committees) backed, such as permitting toxic waste facilities in Black Americans neighborhoods, which intensified slum conditions. This pattern grew for most of the twentieth century across the country. Most recently, the Flint (MI) water issue is one of many examples of damages created by past zoning in specific low-income and racially segregated neighborhoods. According to the Center for Disease Control and Prevention (CDC):

> On April 25, 2014, the City of Flint, Michigan changed their municipal water supply source from the Detroit-supplied Lake Huron water to the Flint River. The switch caused water distribution pipes to corrode and leach lead and other contaminants into municipal drinking water. In October 2016, Flint residents were advised not to drink the municipal tap water unless it had been filtered through a NSF International approved filter certified to remove lead. Although the city reconnected to the original Detroit water system that same month, the potential damage was already done and a state of emergency was declared on January 16, 2016.[7]

[7] https://www.cdc.gov/nceh/casper/pdf-html/flint_water_crisis_pdf.html.

A 2017 report from the Clean Air Task Force (CATF) and the National Association for the Advancement of Colored People (NAACP)[8] documented the specific health risks from airborne pollutants caused by oil and gas development impacting African American communities.

Even though the 1954 landmark Brown v. Board of Education of Topeka Supreme Court ruling against segregation in public schools had a broad influence on other types of segregation, the phenomenon still occurred in practice due to, for example, redlining practices, which ended (at least de jure) with the passage of the Fair Housing Act in 1968. Redlining allowed schooling discrimination via housing. When town officials wanted to keep Black and White communities separated, they segregated communities based on housing so children would attend the closest school to their community. Busing emerged as a partial solution, i.e., transporting a particular child from one race to a school where another race is predominant, in an attempt to promote racial integration. But, even after 1968, and with busing practices being widespread in the US, private schools (especially Catholic schools) attempted to retain the right to admit students on an arbitrary basis but often based on race. In the aftermath of the Brown, the Internal Revenue Service (IRS) attempted to stop this practice by stripping Catholic schools from the right to perpetuate segregation was, however, blocked by the US Senate. An extract of the original New York Times article reporting this news follows.

> WASHINGTON, July 26 (AP)—The Senate Finance Committee declined today to oppose an amendment, passed by the House, that prohibits a crackdown on the tax-exempt status of racially segregated private schools. The amendment, attached to an appropriations bill, prohibits use of Federal funds to carry out the crackdown. The Internal Revenue Service proposed last August a procedure for determining whether certain tax-exempt private schools were discriminating against black pupils. The proposal, aimed at "segregation academies" that were set up to avoid integration, ran into a storm of protest and was revised in February. Bias Against Religion Seen However, some members of Congress feared that the rule could be used, for example, against Jewish or Roman Catholic

[8] https://cdn.catf.us/wp-content/uploads/2017/11/21094509/CATF_Pub_FumesAcrossTheFenceLine.pdf.

schools whose religious constituencies in a given neighborhood happened to be all-white.[9]

This brings to the surface two issues, which are important to understand the problematic role of AI and algorithms in our society. The first is about dysfunctional democracies (the US provides an exemplary case), an issue I unpack further later in this chapter. The second issue is about the importance of government intervention.

Government intervention has proven effective when generating laws regulating markets. In fact, the lack of government oversight has led to market behaviors detrimental to society and that specifically affect people already struggling. Three examples are, in my opinion, paramount to understand the need for government intervention in current times as they demonstrate how the lack of government oversight affects people at the margin at a higher rate. The first example dates from 2007 until 2009 and concerns what goes under the name of Great Recession. Vulnerabilities in the global financial systems surfaced due to the bursting US housing bubble; real estate prices, for years at high peaks, started falling and homeowners began to abandon their mortgages. As a result, mortgage-backed securities declined in value, which caused several to default or require a bailout.[10] Here, two concurrent factors contributed to the recession. Banks were struggling with being able to fund businesses because they lacked sufficient cash flow. Banks' struggle with cash flow and homeowners paying off their mortgages resulted from the decline of the value of the real estate market, so homeowners were paying off debts rather than borrowing money, which lowered the value of mortgage-backed securities. All this was aggravated by longstanding practices of banks granting high-risk, high interest mortgages to people who were not likely to pay off their loans (subprime mortgages practices). Once the market crashed, all these mortgages defaulted, which added the market crisis that soon propagated globally.

Here we need to answer two important questions. First, who were the victims of the Great Recession? Second, how was the Great Recession addressed by institutions? As broadly documented in Rothstein (2017),

[9] https://www.nytimes.com/1979/07/27/archives/irs-plan-on-segregated-private-schools-threatened-bias-against.html.

[10] https://web.archive.org/web/20210626165051/https://www.investopedia.com/terms/g/great-recession.asp.

most "victims" of subprime mortgages were Black Americans. Borrowers needed to better understand conditions and interests of mortgages they were unlikely to repay, but the information provided to them was very poor and, in some instances, misleading and lacking transparency. The problem was that, once the housing market crashed, defaulting borrowers faced foreclose and were evicted. At the same time, the banks, after regaining property of the houses of the defaulted customers, could not sell them at a price that would make up the initial investment in borrowing to their newly evicted previous customers. As an example: a house cost US$100,000. A borrower accesses a loan for US $90,000. The borrower defaults due to high interests and economic crisis. The bank sells the property, which is now worth US$50,000, and ends up with a loss of up to US$40,000. Multiply this for thousand bank loans and add administrative costs to (try to) sell properties. Also, banks needed to sell foreclosed real estate assets to fund businesses. To wit, the banking system collapsed, some bank managers and executives faced indictment, and defaulters became homeless.

Borrowers should have been more careful when accepting loans at very high interests, but the biggest issue was the lack of transparency of the US financial system. Mortgage lenders could persuade borrowers to take loans at subprime rates without disclosing the consequences. This behavior was led by incentives (for borrowers) such as yield spread premiums (YSPs)—basically, rewards for selling mortgages with interest rates higher than these recommended by their financial institutions. Brokers and loan managers literally manipulated potential borrowers to try and hook them with quick closing mortgage deals that they could scarcely repay simply to make short-term profits with commissions. Banks let these things happen with the idea to eventually sell real estates to the market as borrowers defaulted with their houses foreclosed. Often, these infamous deals were offered to Black Americans, historically banned from the real estate market or allowed to purchase homes only in "designated" redlined neighborhoods.

The response to the Great Recession, the Dodd–Frank Wall Street Reform and Consumer Protection Act[11] (known as *Dodd-Frank*), which prevented banks from making overly risky investments (including allowing subprime mortgages, which are investments at the expense of citizens),

[11] https://www.cftc.gov/LawRegulation/DoddFrankAct/index.htm.

in order to prevent future collapses of the banking system. Among other things, Dodd-Frank prohibited incentives for brokers who work for lenders such as the YSPs incentives. The Dodd-Frank act increased the stability of the US (and worldwide) financial system.

The second example very much relates to the Great Recession and concerns the recent collapse of the Silicon Valley Bank (SVB). Before the COVID-19 pandemic and over the past two decades, SVB had targeted relatively few customers, mainly high-tech startups and local companies that needed a considerable amount of cash to run their businesses. Relying on few customers exposed SVB to industry breakdowns. If the high-tech industry slowed down, all of SVB's customers would be affected. But, unfortunately, it is hard for banks to choose their customers—in fact, generally it is the other way around. When the pandemic hit, the global economies slowed but high-tech startups that provided online services proliferated (online grocery shopping, streaming services, social network-based gym equipment, and so on). For instance, during the first three months of the pandemic (March–May, 2020) Amazon increased its revenues of 44% from the same period in the previous year. This and similarly huge profits that "online" companies made generated an overwhelming influx of money into banks such as SVB. SVB decided to invest this money in long-term (10 years) bonds—safe and remunerative investments, especially in regime of low interest rates as in 2022.

Government bonds are sensitive to interest rates. They appreciate if interest rates rise. For instance, if one lends $100 the government (10 years bond) at an interest rate of 1%, one will receive a periodic dividend of 1% and will receive the capital ($100) after 10 years. The value of the bond depreciates if the interests rise. If one year later one does the same thing and the interest rate is 5%, then this second bond will have higher revenues. Because bonds can be sold, a bond with 1% interest would be sold for less than its capital value ($100) in regime of 5% interest. The issue here is that SVB focused mainly on federal bonds and did not diversify its portfolio. SVB was allowed to do so because, in 2017, under the Trump administration, bank regulations (cf. Dodd-Frank) were relaxed in terms of investment portfolio for medium-size banks, such as SVB. Who were the victims of the irresponsible investment practices of SVB? Certainly not big high-tech companies and investors. Instead, the victims were small startups such as Tiffany Dofu, a Black American and the founder of The Cru,[12] a professional training and coaching startup aimed at keeping women in the workforce.[13]

The third example of lack of regulations that enable financial institutions' opportunistic (and erratic) behaviors at the expense of women, minorities, and financially weak populations concerns the 2022 crypto crash: the collapse of cryptomarket that involved companies such as Terra (LUNA) and TerraUSD (UST).[14] In July, 2022, the cryptomarket lost US$600B in just one week.[15] But, here, again, who were the victims? Cryptomarkets are currently highly unregulated, and it is easier to invest in such market rather than in traditional stock markets. This easy access to investments quickly attracted young Black Americans, a population that has historically been discriminated by financial institutions. As a 2022 report from The Atlantic puts forward[16]:

> The prototypical face of crypto is young, white, techy, and male, but perhaps no other demographic group has been harder hit by the crypto bust than Black Americans, who are half as likely to own stocks as their white counterparts but significantly more likely to own cryptocurrencies. Because Black investors piled into the crypto market at or near its most recent top, many of those investors are now in the red. That is especially worrisome because Black investors had so little to lose to begin with: Young Black men are one of the poorest segments of American society. It is also worrisome because many Black investors poured money into bitcoin because they found it so hard to build generational wealth in the first place. Discriminated against by banks, overlooked by investment managers, redlined and saddled with educational debt, many turned to more esoteric opportunities.

This third example too, demonstrates how the lack of regulations penalizes people at the margin in a disproportionate way. All three examples concern the financial industry (even if remains unclear how

[12] https://thecru.com.

[13] https://finance.yahoo.com/news/real-victims-svb-aren-T-201307893.html.

[14] https://www.terra.money

[15] https://www.investopedia.com/crypto-crash-what-investors-need-to-know-5272147.

[16] https://www.theatlantic.com/ideas/archive/2022/11/black-investors-bitcoin-cryptocurrency-crash/671750/.

regulators will position cryptocurrencies with respect to the US financial system). In May 2023, New York State Attorney General James proposed nation-landing regulations on the cryptocurrency industry[17] focusing on protection, transparency, and oversight. It is interesting how history resurfaces longstanding issues in various forms. In the first part of the nineteenth century, despite the Fourteenth Amendment (1868) spelling out equal protection of the laws for all citizens, housing discrimination against Black communities was the norm until the Fair Housing Act, in 1968. In the 2010s, banks exploited Black communities by selling them un-repayable loans until Dodd-Frank kicked in. In more recent times, Dodd-Frank was partially turned over, and the victims, once again, were Black Americans. And the lack of regulations of the cryptocurrency market led to other discriminatory injustices affecting historically penalized and marginalized populations.

My view is that market (vs. state) does not solve problems of social injustice nor does it address discrimination, privacy issues, or, more generally, DEI issues. It didn't address social issues in the aftermath of the Roosevelt's presidency when the Republican Party attacked the New Deal with consequences associated with discrimination, segregation, and so on that the Black community still deals with. For instance, the fact that Black Americans are less likely to earn a college degree than White Americans[18] can be easily linked with past segregation as research showcases that college-educated parents are more likely to have children who follow suit.[19] Schooling segregation determined that Black students attended poorly funded public schools that were less able to prepare students to college compared to "White-based" schools—often private schools with more resources, lower student-to-teacher ratios, and better able to prepare their students for college (Phillips-Fein, 2009). Likewise, redlining practices restricted Black Americans from owning houses, which led to the current trend where young Black Americans inherited debt rather homes.

And, in a similar way, the market will not address social issues associated with the unregulated use of AI systems to hire/fire people, support

[17] https://ag.ny.gov/press-release/2023/attorney-general-james-proposes-nation-leading-regulations-cryptocurrency.

[18] https://www.usatoday.com/story/news/education/2023/05/15/college-student-gap-between-black-white-americans-worse/70195689007/.

[19] https://nces.ed.gov.

the juridical system, and sell products on social media—just to mention three main pervasive and invasive applications of AI. As history shows how market-based policies (lack of government intervention, or government supporting businesses at the expenses of individuals, especially if marginalized) have not worked well for people at the margin, I advocate for strong government intervention and actions to regulate AI and its most recent developments (e.g., generative AI).

One important issue that might negatively affect governments' ability to effective regulating AI systems is the intervention (or power, which translates into lobbying activities) of businesses who aim to make profits from AI, never mind if this is done at the expenses of citizens. In reaction to the New Deal, businesses attempted to influence the government in several ways in the late 1930s. Noticable actions included the creation of business-related associations such as Jouett Shouse' American Liberty League, a political organization formed in 1934 by wealthy business elites and prominent political conservatives who opposed the New Deal and emphasized private property and individual liberties (Rudolph, 1950) and the Heritage Foundation, an US activist conservative think-tank based in Washington that took a leading role in the conservative movement during the Reagan presidency.[20] In fact, these kinds of associations were born as private initiatives by businessmen, with the goal of influencing politicians and their policies, something that now is best practiced with lobbying. In the early 1900s, large businesses such as DuPont invested big money to create associations of activists such as the ones just mentioned. Nowadays, Apple, Amazon, Google, and Facebook spend millions on lobbying the Federal Government according to 2021 data.[21] On the one hand, we need to go past believing that markets can regulate society due to the inherited disadvantage of the Black community, as well as other minorities and people at the margin. On the other hand, it is important to avoid passing over regulations to the state if governments are influenced by the very businesses that were supposed to rule in a market-based regime.

In fact, businesses are by definition revenue-driven. The best case they make is that, if they don't generate revenue, they will generate unemployment because that's all businesses do—focus on attaining sustainability in

[20] https://www.heritage.org.

[21] https://www.politico.com/newsletters/morning-tech/2022/01/24/tech-spent-big-on-lobbying-last-year-00001144.

the long term to pay for capital and labor, despite the fact that CEOs receive 399 times as much more remuneration than a typical worker, as per a 2021 report by the Economic Policy Institute.[22] If laws and regulations are not put in place that limit high-tech's intervention in people's everyday life with widespread AI affecting the workplace, social media, the mortgage market, and so on, private citizens will become test users of disruptive innovations such as generative AI, the long-term consequences of its use being unknown to most. For instance, Eric Emerson Schmidt, Google's CEO from 2011 to 2011 and Executive Chairman from 2011 to 2015, said in 2021: "Why don't we wait until something bad happens and then we can figure out how to regulate it—otherwise, you're going to slow everybody down. Trust me, China is not busy stopping things because of regulation."[23] Schmidt's statement is highly problematic for many reasons.

First, as I pointed out in Chapter 3, unleashing to the public generative AI is not the same thing of "testing" Microsoft Excel because, while the latter might crash a balance sheet, the former can crash families or communities. Second, a line of thought such as that of Mr. Schmidt suggests that time to market—for competition (see his point on China)—is more important than public safety. Here, once again the interests of corporations are put ahead of public wellbeing. Think of the risk of deploying self-driving cars without thorough road in-lab tests. I am sure that Tesla attempted to test their vehicles before putting their cars on public roads. Yet, US laws are very relaxed with respect to self-driving car being tested and approved for safety before being offered to the public. During a speech in 2022, Tesla CEO Elon Musk himself summarized the difference in transatlantic car regulations: "In the U.S. things are legal by default, and in Europe they're illegal by default.[24]" Third, Schmidt's argument seems to imply that lab tests of innovations are insufficient to assess their effectiveness. If my interpretation of Mr. Schmidt's point is correct, why not using human beings rather than dummies when assessing cars' safety during crash tests?

[22] https://www.epi.org/publication/ceo-pay-in-2021.

[23] https://www.youtube.com/watch?v=CmBpsw1ORQ0.

[24] https://slate.com/technology/2023/02/tesla-recall-full-self-driving-nhtsa-musk-regulation.html.

While using "real" people during crash tests is visibly unethical because the harms overcome the benefits (one of the key principles of ethical research), deploying AI systems to society has invisible consequences, because they affect mainly people at the margin who have very few means to be heard. For instance, when the COMPAS algorithm was deployed in the juridical system (cf. Courtland, 2018), suspects awaiting trial who were denied bail were not able to protest because they were in jail and didn't even know that an algorithm was responsible. In addition to that, as the Courtland study suggests, these people were likely to be part of disadvantaged communities and populations (in large part Black Americans). Fourth, there is no evidence of the fact that regulating markets and the deployment of products and services hinders innovation. Mr. Schmidt's point echoes that of former President Reagan who, in 1980, argued that overregulation was letting down the American economy and making it more difficult for American corporations to compete abroad. As such, Reagan supported the Kempt-Roth tax cuts and promised to limit the size of the Federal Government to reduce bureaucracy and avoid unnecessary deficits. But a 2023 report issued by the National Bureau of Economic Research[25] says otherwise: interventions of the Federal Trade Commission (FTC) and the Department of Justice (DOJ) do not hinder innovation. On the contrary, the report reads:

> The quantity and price of output, the increase in production inputs (employment), together with a proportionally smaller increase in sales, strongly suggests that these DOJ antitrust enforcement actions increase the quantity of output and simultaneously decrease the price of output. Our results show that government antitrust enforcement leads to persistently higher levels of economic activity in targeted industries.

But US resistance to impose rules defining the extent to which corporations can (or can't) abuse private citizens in the name of innovation, profits, and worldwide leadership is contrasted by other geopolitical contexts (mainly China and Europe), which I discuss next.

[25] https://www.nber.org/system/files/working_papers/w31597/w31597.pdf.

AI Legislations by Geographical Areas

AI is a global phenomenon like the Internet. But, unlike the Internet, the extent to which AI is developed, regulated, and used is country specific (or "block" specific as per the EU). In this chapter, I have thus far highly criticized the US for its lack of attention to private citizens, especially if they belong to unrepresented communities, and I've focused on Black American particularly. But no system lacks flaws. The US has a history of slavery and segregation as I explain above. This past has materialized into the lack of attention for the very people who were put in slavery conditions and were segregated—Black Americans who, because of the data being collected, put them in a condition of constant surveillance, which resembles structural racism as described by critical race theory (CRT) studies. Other countries are implementing AI-related regulations that reflect their own history, culture, and beliefs.

China has a long history of government domination. As Wikipedia reports:

> The politics of the People's Republic of China takes place in a framework of a unitary Marxist–Leninist one-party socialist state under the Chinese Communist Party (CCP). The Chinese political system is authoritarian. There are no freely elected national leaders, political opposition is suppressed, all religious activity is controlled by the CCP, dissent is not permitted, and civil rights are curtailed. Direct elections occur only at the local level, not the national level, with all candidate nominations controlled by the CCP.[26]

But what might sound strange to some, China has recently built one of the most modern system of data management, accounting for privacy, that mandates all private companies to protect citizens data in a way that outweigh many Western democracies. The Berkeley Office of Ethics[27] does a very good job in laying out the key principles of China' Personal Information Protection Law (PIPL), adopted on August 20, 2021, and effective on November 1 of the same year. Because of the international breadth of the law, it affects all companies dealing with personal data

[26] https://en.wikipedia.org/wiki/politics_of_china.

[27] https://ethics.berkeley.edu/privacy/international-privacy-laws/china-privacy-law.

of Chinese citizens, which means that it is not only applicable to organizations and individuals who process persona data in China but also organizations that process data of China citizens outside the country. For this reason, the Chinese Government has created a website in English that spells out all the key rules of the law.[28] The document is titled "Personal Information Protection Law of the People's Republic of China," which clarifies that the law is about individuals, not borders. This is or will become an issue, for instance, for American companies as it happens now when dealing with GDPR. In October 2023, China's National Information Security Standardization Technical Committee released a draft document that laid out principles concerning how to determine whether a generative AI model is problematic.[29]

The European GDPR is in fact quite similar to PIPL. GDPR is an EU regulation on information privacy that also governs the transfer of EU citizens data outside Europe. GDPR enhances individuals' control over their personal information, such as the right to be forgotten that I mention in in Chapter 2, and spells out that EU citizens, under certain conditions, have the right to have their information be removed from the Internet.[30] Here, an important similarity between PIPL and GDPR with respect to data control is that they both revolve around consent. Once the end user provides consent to data treatment, the data collector (e.g., a social media company or a marketing company) is cleared. Other similarities concern:

1. The breadth of the scope (and coverage/protection) of the laws. Both laws cover data privacy including extraterritorial applications where people's data is recorded or processed outside the borders of the sovereign countries or blocks (China or the EU).
2. The minimization of data collection. Both laws require organizations to collect only the data necessary to provide a service. For instance, a university enrolling students should not ask for data concerning sexual orientation because this data point does not affect academic performance. This stems from the principle that data

[28] http://en.npc.gov.cn.cdurl.cn/2021-12/29/c_694559.htm.

[29] https://www.technologyreview.com/2023/10/18/1081846/generative-ai-safety-censorship-china/.

[30] https://gdpr.eu/right-to-be-forgotten/.

collected should not be used for purposes not related to the reason why the data was collected in the first place. This principle minimizes data reuse, which prevents what I call *data backfire*: when someone agrees to give up their data to a company for a purpose, but then the data is used for other purposes, which can damage the individual. A typical example of this issue, which I often offer to my graduate students, is about healthcare data, which might be collected by smartwatch companies and then used by insurance companies to set higher premium on the basis of preexisting conditions such as high blood pressure or diabetes.

3. The right to access individual data, which includes data correction and deletion (cf. right to be forgotten). Here the issue is that, while GDPR applies to any data (for instance, a wrongful arrest by law enforcement), it is not clear whether PIPL applies to government-related data. In fact, a key, more general, difference between GDPR and PIPL with respect to government (vs. private companies) is that the latter applies only to companies and not to the state, which means that all the privacy rules that apply to companies do not apply to the government, which can access citizen data at any time. But, if we keep the analysis at the company level, both EU and Chinese citizens have the right to learn about what data a company has on them and can ask for their deletion, with reason.
4. The centrality of consent. Both GDPR and PIPL allow companies to collect and use citizen data insofar the data generator agrees with it. Practices associated with lengthy terms and conditions, where data privacy clauses are hidden and written using legal jargons, are still questioned. Yet, as long as companies can obtain consent in some way, they can use people's data. A different issue is how the data can be used. In Europe, a general principle is that the data should be used for the purpose they were collected. From the Chinese PILP, this point is not clear.
5. Data reuse. It is related to minimizing data collection. Both GDPR and PILP recommend that companies collect just the data needed for the purpose of data gathering. For instance, a car insurance company should not collect data such as people's height and weight because they are irrelevant to car/driver safety. This data will be prone to *reuse*, a debatable if not illegal practice. For instance, someone who has to have an accident that doesn't involve other vehicles (say, the driver hits a tree while driving off-road overnight

and becomes injured) should be processed by the insurance companies with respect to road conditions, high speed, whether the driver was intoxicated, and so on, but considering weight as a possible cause of the accident because it is more likely to be associated with sudden illness would not be appropriate (besides the difficulty of proving such claim). Therefore, because a driver's weight is not admittable as an element as a reason for why an accident occurred, such data should not be collected.

In Europe, data reuse was brought to the surface with respect to generative AI (specifically when ChatGPT became available in the bloc) due to the ambiguities associated with how OpenAI aimed to use data collected from prompts generated by users. Is obvious that ChatGPT learned from prompts; when a ChatGPT user asks a question using several phrasings (and attempts), the software learns about the many ways users ask questions and, therefore, refines the way it understands such questions. While the Italy case, where the Italian Data Protection Authority had accused ChatGPT of unlawfully collecting users' data (see Chapter 2), concerned the lack of protection of minors, a more general question is about whether ChatGPT (and other generative AI-powered chatbots such as Bard from Google) users prompt data to feed its training dataset. In other words, is it possible that one's question containing information on a subject are used to answer someone else's question? Here is a clarifying example: a researcher on stem cells asks ChatGPT about a relationship between two proteins based on an intuition that their interactions might lead to a discovery. ChatGPT doesn't know the answer to the question but records in its database that these two proteins might have a relation. Another research asks ChatGPT about potential relations between random proteins to build stem cell models and the generative AI provides an answer that draws on the prompt of the other researcher. While this scenario may seem hypothetical to the most, companies such as Apple have recently restricted employee use of ChatGPT for this very reason. As a 2023 *Wall Street Journal* reports[31]:

[31] https://www.wsj.com/articles/apple-restricts-use-of-chatgpt-joining-other-companies-wary-of-leaks-d44d7d34.

> Apple has restricted the use of ChatGPT and other external artificial intelligence tools for some employees as it develops its own similar technology, according to a document reviewed by The Wall Street Journal and people familiar with the matter. Apple is concerned workers who use these types of programs could release confidential data, according to the document. Apple also told its employees not to use Microsoft-owned GitHub's Copilot, which automates the writing of software code, the document said. ChatGPT, created by Microsoft-backed OpenAI, is a chatbot derived from a so-called large language model that is able to answer questions, write essays and perform other tasks in humanlike ways. When people use these models, data is sent back to the developer to enable continued improvements, presenting the potential for an organization to unintentionally share proprietary or confidential information.

The possibility of privacy breaches is concrete and poses important ethical issues concerning individual privacy and corporate intellectual property. To this end, the principle of minimization of data collection becomes a central issue with generative AI because the data collection (and trade) becomes automated with little oversight by human beings.

As previously mentioned, one caveat is that, while in the EU, sovereign states are upheld to the same standards of private companies, in China, the state can access citizens data with no specific limitations. This is not surprising given the different culture and regimes in the EU and China. However, in defense of the latter, in preventing companies' control over citizens, the Chinese Government can retain oversight over the Chinese population while limiting deceptive actions performed by private companies. From the perspective of a Chinese citizen who believes that the government provides for them with social infrastructures (including laws) limiting or preventing the private sector from collecting personal data might make sense. While Western cultures might not be fine with the broad oversight of Chinese Government over their citizens, I am not making bold judgments on the topic as a form of respect of a culture that I am not part of.

Another important and very recent regulation is EU's AI Act, a proposed regulation that, once again, highlights the privacy rights of human subjects. The AI Act aims to classify and regulate AI applications in organizations and it is based on their risk to cause harm to people. This classification primarily falls into three categories: banned practices, high-risk systems, and other AI systems (cf. also Veale & Zuiderveen Borgesius, 2021). Banned practices concern AI applications that clearly harm people (mainly concerning privacy and discrimination). For instance, AI-equipped cameras that single out people in public places and assess their risk of committing a crime based on the color of their skin (predictive policing) is highly discriminatory (and racist) and is based on historical—flawed—data. High-risk systems concern AI applications that *may* carry benefits for individuals, but there is not yet evidence of whether the benefits overcome harms—a basic principle of responsible research that stems from the 2012 Menlo report[32] whose principles rely on the 1979 Belmont Report[33]

More generally, high-risk systems include all systems that pose threats to the health, safety, or the fundamental rights of people. For instance, innovation unleashed to society with poor vetting—the case of health applications where AI provides ambiguous outputs—may harm people, while the benefits are not clear, if there are benefits at all for end users. These systems include poorly tested medical devices—I wrote a paper concerning innovation irresponsibly deployed during the COVID-19 pandemic (cf. Marabelli et al., 2021)—and other systems that outcomes are ambiguous and likely to unfold potential discriminations or privacy concerns. Other AI systems are not subject to specific regulations. AI systems that are "AI Act exempt" are systems that do not operate close to human subjects, e.g., AI systems that supervise line production of goods in manufacturing companies, which are believed to not pose direct harm to people, even if some might argue that indirect harms are equally relevant. Examples include automation, which might create unemployment, social discontent, protests, and strikes, which could result in economic instability and, ultimately, social injustice and racial discrimination as unemployment generated by automation generally hits people who are poorly educated and financially weak.

The AI Act, if turned into an official EU law, has the potential to become a global standard following the path of the GDPR. I discussed these legal topics with Khiara Bridges, who expressed concern about whether and how fast the US will learn from the EU (or China). She argued that:

[32] https://www.dhs.gov/sites/default/files/publications/CSD-MenloPrinciplesCORE-20120803_1.pdf.

[33] https://www.hhs.gov/ohrp/regulations-and-policy/belmont-report/index.html.

> I think it's easier to follow suit as opposed to innovate. But I think also the US has been very steeped in American exceptionalism and the idea that we can't learn from other places because America is just so unique. This definitely comes up in constitutional law all the time. There's likely a majority of Justices on the [Supreme] Court right now that feels like it doesn't matter what's happening internationally. It doesn't matter what human rights bodies are doing. It doesn't matter how other countries are interpreting their constitutions. The idea is that the US Constitution ought to be completely siloed away from developments and insights from other countries. Many members of Congress, a majority of Congress, probably feel the same way – that we can't just cut and paste what's happening in Europe or China or elsewhere because the American context is unique, exceptional.

I couldn't agree more with this line of thought. It raises questions about the dysfunctional US system that oversees the creation of laws (US Congress) in several aspects concerning the handling of new algorithmic-based technologies and the lack of attention of such system for social injustices.

Dysfunctional Law-making System in the US and Consequences for AI Legislation

In the US, federal and state laws regulating the use of algorithmic-based innovations (AI as a primer) are lacking. In this section, I provide hard evidence of why the US Congress needs to take concrete actions to prevent or at least limit harms to people caused by the widespread use of these technologies, in particular because the harms systematically affect people at the margin who have very poor means to defend themselves (cf. also Eubanks, 2018).

America has changed in the last decade due to the widespread use of social media and the Trump presidency, which have both contributed to high polarization. Yet, recent events (e.g., the murder of George Floyd, #metoo movement, and the Black Lives Matter movement) and social activists (e.g., LGBTQ+ activists) have exacerbated the differences between what some call "woke" ideology (literally meaning "alert to racial prejudice and discrimination"), which now identifies "far leftists" and people with awareness of social injustice who want to give a voice to marginalized people. While here I don't aim to get into political debates concerning woke (or the meaning both left and right wings attribute to this term), it is important to acknowledge woke as a term and a movement to understand how polarization in the US has led to dysfunctional practices at the government level.

The aftermath of the Obama Administration left the country in a place where, for most Republicans, the eight years of the first Black president's administration was something close to socialism—mainly (but not only) due to the bailout of the automotive industry during the Great Recession. Government intervention was huge, especially during President Obama's first mandate. Yet, Obama left office as the "Deporter in Chief" because, under his administration we witnessed the highest number of undocumented people being removed from the country.[34] Nevertheless, the reaction from the Trump Administration (which survived during President Biden's first mandate) was the quintessence of the Republican approach to public intervention in the private sector: nothing. This laissez faire Reganian philosophy has substantially affected the way algorithmic-based technologies have been handled especially with respect to social media and generative AI. Until recently, where (mainly inconclusive) conversations between the US Congress and high-tech companies have taken place with the goal of discussing the potential threats to society posed by AI, nothing has been done to prevent the irresponsible use of this technology.

While I do not delve into the dysfunctional mechanisms of US lawmaking (and, more at large, juridical) system here, I do expand on two key issues that I recently discussed when examining issues associated with AI-based data collection in virtual settings (Marabelli & Newell, 2023). The first issue concerns the lack of whistleblower protection that social media employees have when witnessing misconduct. The second issue concerns research that is undertaken by social media companies, which make intense use of AI and algorithms to force users to spend several hours a day on these platforms (TikTok being the quintessential example).

Whistleblowers are key organizational actors who uncover organizational misconduct. The Office of the Inspector General of the U.S. Department of Justice defines whistleblowers as follows:

> Whistleblowers perform an important service for the public and the Department of Justice (DOJ) when they report evidence of wrongdoing.

[34] https://www.npr.org/2017/01/20/510799842/obama-leaves-office-as-deporter-in-chief.

> All DOJ employees, contractors, subcontractors, grantees, subgrantees, and personal services contractors are protected from retaliation for making a protected disclosure.[35]

Yet, the lack of legal protection of whistleblowers is well known (cf. Chordiya et al., 2020; Gao et al., 2015). This is problematic because whistleblowers who decide to "go public" and disclose misconduct generally lose their job and find it difficult to find another, because companies don't trust them as employees loyal to them. Interestingly, in the past, whistleblowers have been mostly women, which, according to recent analyses, happens (among other reasons) because women in high-tech companies have less power to influence decisions using formalized command chains (Berman & Lundquist, 2022). A case in point about AI and DEI issues concerns Google's firing Timnit Gebru, then lead researcher in the company's AI ethics unit. She was fired for disclosing (in a published paper) an algorithm's flaw that could potentially lead to discriminating outcomes—and Gebru was one of the few Black people with a leading role in AI the company. Gebru was asked by her boss to do not publish the paper but she decided to do otherwise.

An even more explicit example of the important role played by whistleblowers (and how hard it is for whistleblowers to operate in the context of high-tech companies and their associated male-led culture) is that of Frances Haugen, a former Meta employee hired to help protect against election influence on Facebook who said "she had grown frustrated by what she saw as the company's lack of openness about its platforms' potential for harm and unwillingness to address its flaws," according to a Wall Street Journal's report.[36] Ms. Haugen decided to leave the company before coming out with the press, which showcases the lack of support that people have when feeling uncomfortable with ongoing issues in the workplace. In particular, Ms. Haugen reported that internal research at Instagram (a Meta-owned company) found that the algorithm that managed news feeds on the platform offered posts suggesting strict

[35] https://oig.justice.gov/hotline/whistleblower-protection.

[36] https://www.wsj.com/articles/facebook-whistleblower-frances-haugen-says-she-wants-to-fix-the-company-not-harm-it-11633304122.

weight loss diets to teenagers who already had an eating disorder and showed inappropriate posts to teenagers with suicide thoughts.[37]

Where is the problem here? Well, in part, the problem is that most social media platforms are managed by algorithms that "lock in" users because this is a way these platforms sell ads. To wit, they do so by show-casing perspective customers (whoever wants to post ads on social media) that they retain a critical mass of users and that these users spend a lot of time on the platform each day. That then the ads are targeted and that social media companies can do so by mining user data is also true (and drives the prices of ads), but here we focus on lock in strategies, which are made by "appealing" feeds, something TikTok currently masters, in the realm of social media for young generations. But of course no one wants to believe that companies such as Meta would design an algorithm that purposely offers posts that might encourage suicide or exacerbate eating disorders in teenagers.[38] Here the problem is that algorithms are imperfect and need to be managed as Luca et al. (2016) explain with rich examples in their *Harvard Business Review* article. Companies are revenue-driven, and I believe that it will be very naïve to think that they can act responsibly just because they follow ethical principles. The issue put forward by Ms. Haugen is one of many, but we know just a few of them because internal research conducted by companies is seldom made public unless it unveils results that make these companies look good. It is here (along with whistleblower protection) that lawmakers should step in.

Internal research of data-driven companies such as social media companies (but also companies such as Amazon, Zillow, and Waze, to name three that have or might have issue with the way they handle user data) is not made public, yet it is (or should be) a source of meaningful insights helpful to understand potential ethical and DEI problems. As a matter of fact, who owns data and outcomes on potential harms arising from the use of algorithms and AI are the beneficiaries of data-driven insights. This is a conflict of interest because disclosing problematic research might negatively hit a company's bottom line. Governments and institutions that cannot access internal research data are currently struggling with

[37] https://www.businessinsider.com/facebook-knows-data-instagram-eating-disorders-suicidal-thoughts-whistleblower-2021-10.

[38] https://www.wsj.com/articles/facebook-knows-instagram-is-toxic-for-teen-girls-company-documents-show-11631620739.

making laws precisely because it is hard for them to demonstrate specific negative effects of emerging technologies. For instance, in the case of Instagram, had Ms. Haugen not come out to the press, no one would have known about the issues that the platform faced. In such a system, we can only rely on companies' conscience and need to be confident that they will address problems as they face them, even if limiting algorithms in some ways might diminish user engagement. There are exceptions, and workarounds. For instance, in 2021, the office of US Senator Richard Blumenthal created a fake Instagram profile of a 13-year old girl with the goal to research how the platform might affect teenagers' mental health. According to the senator:

> Our research has shown, in real time, Instagram's recommendations will still latch on to a person's insecurities, a young woman's vulnerabilities about their bodies and drag them into dark places that glorify eating disorders and self-harm.[39]

This finding is astonishing and demonstrates the urgency, for institutions, to access companies' internal research or at least to access the data to pursue research as Research Fellow Gabriel Nicholas at the Center of Democracy & Technologies argued in a 2022 post.[40] Here the paradox is: how can we require that lawmakers regulate a field where they don't have hard evidence of misconduct? The fact that evidence of misconduct cannot be investigated because of the lack of data becomes the relevant problem. While the US is in an impasse with the Senate having frequent closed doors conversations with high-tech companies on how to regulate the most cutting-edge technologies such as generative AI,[41] Europe has taken concrete actions to mandate transparency and accountability. This means that it is possible to address (at least some) problems associated with the widespread use of algorithms and AI. But as Khiara Bridges told me in an interview for this book: "we can't learn from other places because America is just so unique."

[39] https://www.nbcnews.com/news/us-news/blumenthal-S-office-posed-girl-fake-instagram-account-study-app-n1280478.

[40] https://cdt.org/insights/social-media-companies-should-give-researchers-access-to-more-data-other-industries-can-show-them-how/

[41] https://www.bloomberg.com/news/articles/2023-09-13/tech-leaders-head-to-capitol-hill-to-propose-their-own-ai-rules#xj4y7vzkg.

As the 2024 US presidential elections approach, other type of social media data are required to assess how algorithms offer recommendations in users' feeds that concern fake news or otherwise feeds that are generated by, for instance, generative AI-powered bots. In 2022, Twitter (now X) was the only major social media platform that allowed researchers to freely access platform data. But, since 2023, it has revoked that access.[42] The US is experiencing challenges associated with making laws disciplining online algorithms and not having enough data to expose misconduct or at least hard evidence that showcases that these algorithms generate harms (from targeting teenagers, to unfairly influencing election processes). In contrast, EU's 2022 Digital Service Act[43] ensures that large social media platforms (with at least 45 million users) open up relevant data to selected researchers deemed to be independent of commercial interests. Companies can still refuse to give researchers access to their data but at least the Digital Service Act is a top-down, institutional initiative in the right direction.[44]

Meanwhile, with institutional inertia going on, in the US, it is the Supreme Court that leads in matters of AI regulations as recent examples such as the Gonzalez vs. Google ruling demonstrate. But, in the US, it is not uncommon to witness lawsuits yielding results before other forms of regulations kick in. The US approach is diametrical different from, for instance, that taken by the EU, where preventive laws are set in stone to avoid innovation being unleashed in society with unknown outcomes. The US has a different philosophy: wait for harms to emerge, which will be then addressed in a more reactive fashion. In an *MIT Technology Review* article, Melissa Heikkilä reports Amir Ghavi, Partner at the law firm Fried Frank, as stating: "That's a pro-capitalist stance. It fosters innovation. It gives creators and inventors the freedom to be a bit more bold in imagining new solutions.[45]" As I pointed out earlier in the

[42] It is fair to note that Meta has lately release a "content library" app that allows selected researchers to access social media data in real time (https://developers.facebook.com/docs/content-library-api). This might be a step in the right direction, with social media companies deciding to "open up" their research data to third parties such as universities and independent researchers, along with government institutions.

[43] https://digital-strategy.ec.europa.eu/en/policies/digital-services-act-package.

[44] https://www.nature.com/articles/d41586-023-03102-6.

[45] https://www.technologyreview.com/2023/07/17/1076416/judges-lawsuits-dictate-ai-rules/

book, this is America's line of thought. Too many regulations prevent innovation. It might be true, but this is also made at the expense of vulnerable individuals. In the case of Instagram, it is young teenager with mental issues that are discriminated against. In addition, these teenagers let alone with their smartphones are more likely to be children of hard-working parents with little time to care for them. That's not neglecting care. These are situations where parents, to make ends meet, need to work two or three jobs and their children are left with their problems that social media algorithms magnify. Needless to say, these situations are often typical of low-income marginalized populations. But here generalizations are inappropriate. Eating disorders—and many other issues—can hit any individual, and it is just unethical to make profits on them never mind if the out-of-control algorithm didn't mean to create harms.

Circling back to the whistleblower protection (or the lack thereof), one interesting initiative, first pursued by Facebook (now Meta), was to institutionalize the role of oversight boards. These boards should be formed by independent field experts able to rule with respect to innovation to be unleashed to society and to rule in favor or against current practices pursued by the companies for which they advise. These boards should be independent (i.e., with no conflict of interests with the companies they serve). In a 2022 article on Proto,[46] Robin Donovan recalled:

> There's a joke in Silicon Valley about how AI was developed: Privileged coders were building machine learning algorithms to replace their own doting parents with apps that deliver their meals, drive them to work, automate their shopping, manage their schedules, and tuck them in at bedtime. As whimsical as that may sound, AI-driven services often target a demographic that mirrors its creators: white, male workers with little time and more disposable income than they know what to do with.

Here, the point concerning boards is about the need to separate out the people who create AI systems and the people who vet them before these systems go live. But boards have issue. For instance, the Facebook external oversight board, formed by academics and industry experts on ethics matters, was supposed to help the company make difficult ethical decisions with respect to freedom of speech and online safety. Yet, precisely because this is an external board, Facebook did not have to

[46] https://proto.life/2022/06/what-will-it-take-to-decolonize-artificial-intelligence/

follow its recommendations. In sum, oversight boards can act as collective whistleblowers but have been so far rather ineffective because their recommendations are not enforceable.[47]

The Power of Private Companies and the Free Market Issue

Nowadays, the pace at which technologies develop and are unleashed in society is much faster than it used to be even a decade ago when big data became a widespread phenomenon due to an increase in computing capabilities (processing, storage, Internet/networking). In 2023, we can stream high-quality video feeds using a smartphone with the associated potential for data collection, mining, and so on.

It is true that, when cutting-edge technologies were developed by governments rather than private companies, it was easier to regulate them. When the Internet (former Arpanet) was created, Dr. Jon Postel, late Professor at the University of California Los Angeles (UCLA) and the University of Southern California (USC) and considered one of the fathers of Internet, substantially contributed to the design of domain name server (DNS) system—something that still nowadays allows us to reach websites by typing their domain name rather than their IP address on the address bar of our browser. Put in very simple terms, DNS converts the "real" name of a computer in a network (the IP address) into a domain name, which is easier to remember. In 1998, Dr. Postel, Director at Internet Assigned Numbers Authority (IANA) (which granted IP addresses to companies and private citizens) ran a test in protest of Network Solution, a US Government-backed company than had begun to take control of domain names. Network Solutions was then authorized by the National Science Foundation (NSF) to sell domain names at US$50 for the first 30 years per year for the first 2 years. The test (that goes by the name of "DNS root authority test") comprised the following steps: Dr. Postel, emailed 8 (among the 12 total) operators of the Internet's regional root name servers on his own authority and instructed them to reconfigure their DND servers and change the root zone server from Network Solutions to IANA. The operators complied

47 https://www.wired.com/story/inside-metas-oversight-board-two-years-of-pushing-limits/

with Postel's instructions, which divided control over Internet naming between the non-government operators with IANA and the 4 remaining US Government DNS servers at NASA, the Department of Defense, and the Ballistic Research Laboratory. Even if Internet services were never interrupted, he soon received orders from senior government officials to undo this change, which he did.[48]

This is an example of the power and authority of US Government over innovations, such as Internet, which was already widespread in 1998. These kinds of situations are nowadays unthinkable. Electric vehicle technologies, satellite equipment, self-driving cars, generative AI, and precision medicine (to mention a few) are all technologies in the hands of private companies. The potential harms of these technologies are known to the most. Yet, the US Government struggles with disciplining their responsible use.

One problem I've examined in this chapter concerns the coming back of a laissez faire philosophy, which was proven detrimental to fairness during the 1900s in the US (fair housing, unions) because of the idiosyncratic condition of disadvantage of some parts of the population (minorities, protected populations, etc.). Market rules (such as what Milton Friedman advocated for) stand for the principle of equality (not equity) and overlook historical events that make equality an unfair principle. Equality implies that everyone (right now) has the same chances to be successful (or to fail) because the American (social) system offer equal opportunities to whoever wants to succeed. NOT TRUE. In fact, offering equal opportunities to all (kind of) people in America would neglect the past. For instance, because of slavery's legacy, the US Constitution gives Black Americans a special degree of protection. The Thirteenth, Fourteenth, and Fifteenth Amendment, adopted after the Civil War, were specifically designed to make sure that Black American had equal status—in contrast to giving equal opportunities, because that would not consider how history has placed some people in disadvantaged conditions.

This means, for instance, that historical conditions of Black Americans such as those preventing them from purchasing houses during most of the past century (e.g., redlining and the de jure discriminating actions taken by the FHA) need to be addressed by current loan practices. Because Black American were historically not allowed to purchase houses because

[48] https://en.wikipedia.org/wiki/jon_postel.

they could not benefit from government-backed mortgages, now young Black Americans have lesser chances to make ends meet if they lose their job because they will most likely have to pay rent as most don't own a house. Will this mean that Black American should not be fired because of historical reasons? That would be too much and would be an unnecessary generalization—some Black Americans are very privileged, and attributing an income (or wealth) on the basis of the color of the skin would be a racist classification in the first place. But given the power that AI systems have now, it is the government's responsibly to mitigate side effects (especially those that target weak populations) in part for historical reasons.

For instance, offering ads to purchase houses to Whites while offering similar ads for renting houses to Blacks (as Facebook has done) is unfair and has to be regulated, and it is very unlikely that "the market" (i.e., competition between social media companies if there is one) will prevent these racially biased ads from happening. As I suggest in Chapter 4, social media companies are not fully responsible for their "racists" ads (cf. tradeoff between biases algorithms and privacy). Nevertheless, US institutions should regulate the ads market for historical reasons. It might be true that Black Americans look mainly to rent rather than buy, but the past is responsible for this. For instance, during most of the first part of the last century and in some cases until the 1970s, Black workers were banned from several public jobs along with jobs in private companies, a problem that unions played a part in creating. For instance, the construction trade of the 1950s and 1960s largely excluded Black Americans, who could not benefit from home and highway construction booms in the postwar years. Black Americans could not share with Whites the substantial income gains stemming from the two big wage growth periods of these years. In addition, because of the housing supply restrictions for African American outside ghettos, these populations would either purchase very expensive houses or houses whose value would never increase. As Rothstein (2017, p. 167) noted:

> The unconstitutional treatment of African Americans called for remediation, if not during the war, then afterward, yet the Federal Government's complicity in the suppression of African American labor rights and opportunities was never addressed, nor then and not since.

It is now time for the US Government to acknowledge these issues explicitly by taking responsibility, and actions. Given that most AI and algorithmic-based systems embed biases and these biases systematically hit certain (historically disadvantaged) populations, the government (and regulations) rather than the market (laissez faire approach) should be the main principle leading to social justice.

Beryl Lipton, an investigative researcher who works for the Electronic Frontier Foundation,[49] echoed my concerns about laissez faire practices when we discussed data brokers who capture plates on highways and sell datasets to law enforcement. Beryl has longstanding experience in investigating law enforcement surveillance technologies. She made me reflect on the fact that we should see the market vs. government tension as a multidimensional construct. The market includes companies and individuals, which represents a tension per se. In fact, private companies mine data from individuals, most of which cannot prevent it from happening. The government encompasses central institutions (in the US, the Federal Government and the Constitution) and local governments, such as local law enforcement entities (state and county police). I asked Beryl to tell me more about plate scanning systems work in the US:

> Data, historically has been driven by the advertising market. But a lot of that information that is used in advertising has a lot of appeal for government agencies and also for a lot of other people. I think a big concern that I have generally with some of these technologies is the lack of regulation around the secondary data market and the resale of information that either individuals consented to providing initially, but not subsequently to multiple different parties, and also the way, just as you mentioned, that license plate readers can be installed and owned by private entities working on behalf of law enforcement, but then they have ownership over that larger data set that they would never have been able to gather if they didn't have the buy-in of policing agencies. I think the government, and law enforcement in particular over the last couple of decades, certainly since 9/11, have been working very closely with private industry to try to expand the surveillance capabilities. I think that I have a lot of concerns around how that's working because of jobs, but also just because of general freedom and consumer rights reasons.

[49] https://www.eff.org/about/staff/beryl-lipton.

It is precisely the lack of regulations of secondary markets (data brokers and local law enforcements) that represents an issue that the central government needs to address. This vacuum in federal regulations allows for data reuse as I point out earlier in this chapter. With respect to plate scanning systems, Beryl explained that:

> Generally, how it works is that law enforcement agencies will purchase automated license plate reader cameras, or they might purchase something that is ALPR but serves maybe a slightly different purpose. For example, you'll see those speed limits boards or whatever that track your speed limit. But those also have licensed plate readers inside of them. They serve a dual purpose. I think one of the most prominent uses of a lot of licensed plate readers is in easy passes and that thing where previously before you have the little box or even in some situations where you do have the little box, it still scans your license plate and it'll tell you, Oh, you don't have an easy pass. That's fine. We'll find you anyway and send you the bill. Just drive through. Those are all... And I think a lot of... So generally what happens is law enforcement will buy either stationary cameras or cameras that they can attach to their vehicles. And so you'll see this also in a lot of traffic monitoring. So maybe sometimes you'll get a parking ticket in a place where you didn't see an officer, but they come back and they find you later on at a different place and give you a ticket for that earlier place.

What Beryl describes above is, in my opinion, the quintessence of surveillance capitalism (Zuboff, 2019) with people in power (in this case, local law enforcement agencies) acting in an extremely unregulated way and using citizens' personal data to prosecute. The issue here is that fines are not neutral. Mass surveillance hits people at the margin more because a parking ticket is the same amount for an Amazon worker paid minimum wage and a millionaire's Porsche owner. And, still according to Beryl, there is little hope that local law enforcement will cease using these data-powered technologies unless they become regulated. She said:

> Law enforcement is very interested in using new technology as a whole. They like new equipment. Particularly right now, they feel understaffed and very stressed about…. That's how they make their pitches to buy more of this technology…[because] the private surveillance market is…a lot of companies know that law enforcement is a place to get money. They're very actively marketing to police departments all the time. They go to police conferences.

Mass surveillance may also have the unpleasant consequence of perpetuating segregation due to what I call ***modern redlining***, which stems from the creation of "safe places" (whether fancy neighborhoods or smart cities) where people LOVE to be surveilled because they know that other, stereotyped bad actors will take responsibility for crimes (e.g., see Amazon Ring's racist consequences).[50]

Baryl continued:

> There's a big push by the business community to install more surveillance. If we think about Smart Cities, Smart Cities love surveillance. In some cases, that is how certain cities get grants and funding for Smart City initiatives, is to say, We've put up more cameras. Our downtown area is secure. Please bring your businesses here.

A recent conversation I had with Dr. Donna Maria Blancero, Professor of Management and DEI at Bentley University, made me reflect even more on the importance of government intervention in DEI matters. Below is an extract of our conversation on equity/equality and the systemic issues affecting the US.

> But the issue is, I think the bigger issue for me is that we have to change systems and not people. That's really what justice gets us at. It's not a matter of, oh, these poor Black students or these poor women or this group of people just didn't try very hard. No, none of that. It's not the individual level. It's the system level. We didn't build the systems for them to be successful. You think about neighborhoods—we can even talk about neighborhoods in the Boston area. There are some neighborhoods that are very wealthy that tend to be overwhelmingly white. Then we have neighborhoods that are overwhelmingly black or Latino, and the resources are different. When somebody comes as a student to apply to a university…it's not a matter of saying, oh, they're not very good. It's a matter of saying they were not given the options that people in more richer areas were given. It's really not trying to change the person, like not saying, we'll give you more tutoring or we'll put you in a special class. Yeah, that's a short-term thing, I get that. But the bigger issue is to try to change the system. … A very simple example is … do you ever wonder why kitchen cabinets are so high up and so difficult to get to? Because they were built

[50] https://www.cbc.ca/cbcdocspov/features/smart-doorbell-cameras-can-lead-to-racial-profiling.

with men in mind who could reach them. So in the United States, the average woman is 5'4", and the average man is 5'10".

The examples put forward by Dr. Blancero reflect, as well, historical reasons why we cannot assume that a free market approach (vs. a government approach) will address DEI issues. Market principles rely on the ethical behavior of large companies—something that we all know hasn't happened (at least if we refer to the last two decades, social media companies, etc.) due to the hypercompetitive environment that the high-tech industry operates in. Today, the currency—and what high-tech companies trade—is data, which involves discrimination and privacy issues. The free market is certainly productive in the short-term, but for whom? And what will be the long-term consequences of continuing to marginalize already hurt and penalized populations? More generally, what will be the long-term harms to individuals whose data is managed irresponsibly because of lack of regulations? For instance, HIPAA doesn't apply to most companies managing apps that provide health advice. With the proliferation of AI-equipped medical devices, regulators' work has become even more challenging.

In addition to all this, other parts of the world (e.g., the EU) are taking a substantially different stance with respect to how they handle high-tech companies. As we live in a global world, it will be hard for the US to keep protecting large companies that are currently (and systematically) fined by the EU.[51] To wit, free market is a convenient theory for the privileged. But here I stand with Dr. Blancero's line of thought that it is of paramount importance to shift the focus from the individual to the system.

Conclusions

In this chapter, I stress the importance for government oversight over innovations, especially concerning automated systems such as AI systems that are used by private companies (and sometimes by state organizations such as the juridical system) to make decisions on and about people. While I focus mainly on the US, where problems associated with deregulations are more evident than in other regions such as the EU, regulating automated systems is a global problem. For instance, how can

[51] https://www.investopedia.com/biggest-eu-antitrust-fines-against-big-tech-7501089.

a global company be fair with their employees if they can monitor them in one (surveillance-based) way in the US and can't do the same with employees working in Europe because Europe does not allow certain surveillance types? This is a people analytics issue (see Chapter 2) but also something that brings to the surface the inconsistent regulations on technology use across countries. The use of ChatGPT in some countries but not in others might compromise the competitive advantage of some companies. For instance, two companies providing customer support via online chat in the US and in Europe might have to use generative AI in different ways because EU regulations could prevent the company established in Europe from fully exploit the technology's potential, which would affect its bottom line, since, for example, it might have to pay for "real" people to answer customer inquiries. The difference in laws and regulations across countries does not only compromise the competitive environment—it also tells a story about how it is possible to use top-down regulations to limit negative effects (e.g., discriminations) of automated systems. For instance, as I argue at the outset of this chapter, I no longer believe that it is impossible prevent unintended uses of technologies. Maybe it is not possible to design ethical technologies. But it is certainly possible to limit the use of ambiguous technologies in specific contexts where such systems may have a chance to do the most damage. And I provide hard evidence of the fact that regulating technologies does not hinder innovation, even in the US. Specifically concerning the US, what is scary is that history (of discrimination, segregation, etc.) has not served as lessons learned, and the US Government's institutional inertia is perpetuating longstanding segregation systems that simply shifted from de jure to de facto practices.

References

Berg, J. A., & Woods, N. F. (2023). Overturning Roe V. Wade: Consequences for Midlife Women's Health and Well-Being. *Women's Midlife Health, 9*(1), 1–6.

Berman, F., & Lundquist, J. (2022). *Why So Many Tech Whistleblowers are Women*. Fast Company, available at https://www.fastcompany.com/90758593/here-is-why-are-so-many-big-tech-whistleblowers-are-women

Chordiya, R., Sabharwal, M., Relly, J. E., & Berman, E. M. (2020). Organizational Protection for Whistleblowers: A Cross-National Study. *Public Management Review, 22*(4), 527–552.

Courtland, R. (2018). The Bias Detectives: As Machine Learning Infiltrates Society, Scientists Grapple with How to Make Algorithms Fair. *Nature, 558*, 357–360.

Desmond, M. (2023). *Poverty, by America*. New York, Random House Large Print.

Eubanks, V. (2018). *Automating Inequality: How High-Tech Tools Profile, Police, and Punish the Poor*. St. Martin's Press.

Gao, J., Greenberg, R., & Wong-On-Wing, B. (2015). Whistleblowing Intentions of Lower-Level Employees: The Effect of Reporting Channel, Bystanders, and Wrongdoer Power Status. *Journal of Business Ethics, 126*(1), 85–99.

Hanchett, T. W. (2000). The Other 'Subsidized Housing': Federal Aid to Suburbanization, 1940s–1960s. In *From Tenements to the Taylor Homes: In Search of an Urban Housing Policy in Twentieth-Century America* (pp. 163–179). Pennsylvania State University Press.

Levy, M. S. (2016). Holding the FBI Accountable for Hacking the Apple's Software under the Takings Clause. *American University Law Review, 66*(5), 1293–1321.

Luca, M., Kleinberg, J., & Mullainathan, S. (2016). Algorithms Need Managers, Too. *Harvard Business Review, 94*(1), 96–101.

Marabelli, M., Hansen, S., Newell, S., & Frigerio, C. (2017). The Light and Dark Side of the Black Box: Sensor-Based Technology in the Automotive Industry. *Communication of the AIS, 40*(16), 351–374.

Marabelli, M., & Newell, S. (2023). Responsibly Strategizing with the Metaverse: Business Implications and DEI Opportunities and Challenges. *The Journal of Strategic Information Systems, 32*(2), 1–16.

Marabelli, M., Vaast, E., & Li, L. (2021). Preventing Digital Scars of Covid-19. *European Journal of Information Systems, 30*(2), 176–192.

Pekkanen, S. M. (2019). Governing the New Space Race. *American Journal of International Law, 113*, 92–97.

Phillips-Fein, K. (2009). *Invisible Hands: The Businessmen's Crusade against the New Deal*. W. W. Norton.

Rothstein, R. (2017). *The Color of Law: A Forgotten History of How Our Government Segregated America*. Liveright Publishing Corp.

Rudolph, F. (1950). The American Liberty League, 1934–1940. *The American Historical Review, 56*(1), 19–33.

Veale, M., & Zuiderveen Borgesius, F. (2021). Demystifying the Draft Eu Artificial Intelligence Act—Analysing the Good, the Bad, and the Unclear

Elements of the Proposed Approach. *Computer Law Review International, 22*(4), 97–112.
Zegart, A. B. (2022). *American Spy Agencies Are Struggling in the Age of Data.* Wired, available at https://www.wired.com/story/spies-algorithms-artificial-intelligence-cybersecurity-data/.
Zuboff, S. (2019). *The Age of Surveillance Capitalism: The Fight for a Human Future at the New Frontier of Powe*. PublicAffairs.

CHAPTER 6

New Frontiers of AI and Algorithms

In the last decade, we have witnessed a technological revolution that is hardly comparable to any such event from the recent past except perhaps for the birth of Internet and the World Wide Web (WWW) in the late 1980s. But this technological revolution is different than the Internet and, in my opinion, is better comparable to the industrial revolution because it affects people and their rights—the Internet did not at least until Facebook introduced the first News Feeds algorithm in 2009. Calling recent technological advances an AI and algorithms a revolution would not be appropriate because both AI and algorithms have existed for decades. But the combination of AI and algorithms with big data and an exponential increase in processing and storage capacity (including the ability to process and storage personal data) has created the perfect storm for what is happening now—a substantial change of the landscape surrounding our everyday lives. Sensor technologies (something also not new) are becoming more powerful because of how the data being captured is processed in real time and because of AI's ability to make inferences about the outcomes of data processing. All this has the potential to generate immense positive outcomes. Yet, as per what I've been saying throughout the book, classification processes undertaken by AI and algorithms are prone to ethical and DEI issues.

M. Marabelli, *AI, Ethics, and Discrimination in Business*, Palgrave Studies in Equity, Diversity, Inclusion, and Indigenization in Business,
https://doi.org/10.1007/978-3-031-53919-0_6

In this last chapter, with reference to one of my most recent works on emerging technologies that rely on AI and algorithms, I first examine virtual, augmented, and mixed reality platforms with a particular focus on the metaverse and what is/will be the part played by generative AI. Then, I examine other aspects of how AI and algorithms are embedded in existing technologies and settings: robotics, healthcare systems, and the music industry, which serve to illustrate the ambivalent role of such technologies and the potential to negatively affect DEI.

Remotization and Alternative Realities

In the post-COVID-19 era, people throughout the world have become progressively more accustomed to virtual interactions. In fact, the COVID-19 era demanded it. In March 2020, within a matter of weeks, most knowledge workers had to work from home; schools shut down (and reopened online only within weeks); and bars, restaurants, and most nonessential shops closed, even if for periods that varied by country. For instance, in the US we had relatively relaxed rules from our 2020 summer. Europe and China in particular had much strict rules. Essential workers (hospital personnel, employees in pharmacies, supermarkets, bus drivers, among others) were the only ones who did not experience the then-new normality (i.e., working, studying, and interacting with family and friends remotely). Needless to say, in moving their operations online, companies and schools exacerbated longstanding differences between wealthy people and the poor. With respect to businesses, knowledge workers, already better paid than task workers, could work remotely while safe in their homes, which suddenly turned into their office. In contrast, task workers such as supermarket cashiers and bus drivers had to continue to go to work for the same wage and risk contracting COVID. Because it is well-known that wage is a proxy for gender and race (at least in the US), the COVID-19 pandemic amplified racial and gender discriminations with respect to safety. Some press outlets even advanced theories such as that Black Americans were more prone to contract the virus, which ignored the fact that Black Americans were simply more exposed to it. But these racist conjectures and speculations are not new to conversations around access to care. For instance, a 2021 report from the Population Reference Bureau (PRB) notes how Black women in the US are over three times more likely to die in pregnancy and postpartum than White women.[1] This result can be interpreted as a biological issue—something that current science questions (cf. Bridges, 2011)—or as a social issue: Black women in America have less access to care during their lifespan (so are overall less healthy); are, on average, more prone to being overweight, which represents a risk during pregnancy—again, due to the fact that they are less educated and have less financial resources, which leads to unhealthy eating habits; for example, a meal at McDonald's is much cheaper than organic food purchased at a fancy supermarket and prepared and cooked in an equipped kitchen let alone the fact that often people with financial struggles work two or three jobs and have no time to cook; and receive lower-quality care from public hospitals when giving birth.

With respect to schooling, in the initial phase of the pandemic, private schools equipped students with technologies to allow them to attend classes online. Parents of students enrolled in private schools are generally well-educated. This happens because they have good jobs to make enough money to send their children to private schools; thus, these parents could support their children when they struggled with remote learning. All this didn't happen in public schools: parents struggled with needing to stay home with their children but also needing to attend work (e.g., at a local supermarket). As a consequence, women often decided to leave their job as someone needed to stay at home with the children—and women, on average, receive less pay than men, so men often kept their job.

In summary, the pandemic was as bad at killing people as it was at making already struggling populations at the margin more miserable. We will see consequences from the pandemic and the associated lockdowns in the future; for example, children from wealthy families will have fewer learning issues, will be able to score better at tests, and eventually gain admission to prestigious colleges and access prestigious jobs. I describe all the pandemics' ethical and DEI ramifications for disadvantaged populations in more depth in a recent article published in the *European Journal of Information Systems* (Marabelli et al., 2021).

However, the pandemic, if we want to see its bright side, served as an unintended natural experiment for learning more about what remote interactions can benefit DEI, something that I discuss broadly in Marabelli et al. (2023). For instance, as I mention in Chapter 2, due to

[1] https://www.prb.org/resources/black-women-over-three-times-more-likely-to-die-in-pregnancy-postpartum-than-white-women-new-research-finds/.

people being forced to interact online, many academics from the Global South could access prestigious conferences in 2021 to 2022 when many countries went into partial or full lockdown. Online and hybrid scientific conferences do not automatically engender diversity, equity, and inclusion. Offering people the opportunity to attend conferences remotely and, thereby, waiving travel and lodging expenses is a way to broaden access to scientific knowledge for people who would otherwise not be able to afford travel (due to financial struggles, religious beliefs, or their sexual orientation or political orientation). Additionally, women in academia have historically been penalized because they often act as the primary caregiver for children, which limits their ability to travel. With respect to gender and conferences, I ran the 2022 doctoral consortium of the Communication, Digital Technology, and Organization (CTO),[2] a division of the Academy of Management,[3] in hybrid mode. While most attendees physically attended the conference in Seattle, some decided to attend online, such as a pregnant woman who would not be able to attend otherwise and two people from the Global South.

Even though several organizations have reinstated in-person only operations, such as higher education institutions for which dorms often represent a relevant part of their revenue stream, we need to accept that remote access to schooling, work, healthcare, and so on has become more common than before the pandemic. In June 2023, Forbes[4] reported some interesting statistics: in the US, 12.7% of full-time employees worked from home while 28.2% worked in a hybrid model (both at home and in the office), 98% of workers wanted to work remotely at least some of the time, 93% of employers planned to continue conducting job interviews remotely, and 16% of companies operated fully remotely. If we factor in industries where most workers cannot operate remotely such as law enforcement, healthcare, transportation, and brick-and-mortar retailers, one can draw the conclusion that people pretty much enjoy avoiding a commute (also a very green approach to the environment) and are comfortable with working at home. Yet, questions arise concerning people's ability to develop informal networks (e.g., small talk by coffee

[2] https://cto.aom.org.

[3] https://aom.org.

[4] https://www.forbes.com/advisor/business/remote-work-statistics.

machines), which research has proven to promote creativity and innovation (Labianca & Brass, 2006). A recent *Nature* study demonstrated that creativity decreases with Zoom versus in-person brainstorming meetings because the former limits vision to video. Brucks and Levav (2022, p. 1) argue that "this physical difference in shared space compels virtual communicators to narrow their visual field by concentrating on the screen and filtering out peripheral visual stimuli that are not visible or relevant to their partner." Another even more recent paper, still in Nature, suggests that in-person teams make more breakthroughs than remote teams (Lin et al., 2023).

Similar issues apply to remote education. Remote/distance learning diffused way before the pandemic occurred. For instance, the Southern New Hampshire University had over 135,000 online students in 2020 before the pandemic emerged.[5] Yet, attending classes in person promotes peer networking and is paramount in classes that involve hands-on exercises with software, such as in the science, technology, engineering, and mathematics (STEM) disciplines. Note that men mainly pursue STEM degrees, which may raise an issue associated with men more able to attend in-person classes and, thereby, feeds into the current trend of poor women representation in high-tech companies (so-called pipeline effect, see Chapter 3).

The pandemic also significantly affected the healthcare domain. For instance, in the US, as of June 2021, telehealth services had been used 38 times more often than before the start of the pandemic (Bestsennyy et al., 2021), and 95% of healthcare facilities served patients online against only 43% that did before the pandemic (Marr, 2022). While healthcare workers need to operate mostly in person, the healthcare system can benefit from telecare (e.g., it can spread to rural areas where people can find it hard to reach clinics). But what is the quality of telehealth compared to in-person visits? And what are the risks of prescribing drugs to patients without a physical examination? In this book, I broadly discuss problems concerning access to care for communities at the margin, such as Black communities, due to historical reasons. Overall, Zoom is not enough for remote work, education, and healthcare. Zoom meetings prevent people from building informal networks. Online learning doesn't allow class exercises, not to mention disciplines when laboratory classes are central to education, such

[5] https://www.forbes.com/sites/dereknewton/2022/10/20/blip-in-online-college-enrollment-doesnt-mean-what-you-think.

as chemistry, biology, and so on. Telehealth can also deliver subpar care because healthcare professionals can miss patient information via physical examinations.

Virtual, augmented, and mixed realities seem to be the solution to remote interactions. Virtual reality is a simulated experience where users employ wearable devices (headsets to enhance 3D vision along with sensors such as bodysuits). Users are projected in immersive, more-or-less realistic scenarios with which they can interact. Users can also interact with each other and are represented by avatars. Augmented reality is different from virtual reality in that it overlays digital elements on the real world. For instance, with virtual reality, users can experience being on the International Space Station or on a helicopter visiting the Grand Canyon while sitting in their living room. With augmented reality, users can walk in Central Park and see snow on trees during summer. To wit, virtual reality is something that is detached from the real world, while augmented reality needs the real world to work. Also, augmented reality does not necessarily need headsets (or 3D vision). For instance, Pokémon Go is a good example of augmented reality that 3D does not enhance. IKEA has an augmented reality app that allows perspective customers to see what furniture will look like in their homes by placing virtual versions onto their living spaces. Mixed reality overlays digital elements onto the physical/real world but the physical and digital elements interact. Similar to augmented reality, mixed reality is tied to the real world; that is, it will not remove users from their surroundings (e.g., the living room, or Central Park). Yet, users need 3D vision to access mixed reality and can interact with its physical and virtual elements. While Meta has focused on virtual reality with the metaverse, Apple[6] and Microsoft[7] are currently developing mixed reality platforms along with metaverse-like ones. But as per all technologies, tradeoffs surface, which involve accessibility of such platforms, privacy concerns, and discriminations based on digital profiling of race/gender (e.g., online harassment). These issues happen on social media too, but in the metaverse they are magnified by the more realistic impact that these practices have in such immersive environments.

The metaverse is a fascinating world that, in my opinion—as I state in a recent paper with Sue Newell (Marabelli & Newell, 2023) from which I

[6] https://www.apple.com/apple-vision-pro/.

[7] https://www.microsoft.com/en-us/mixed-reality/windows-mixed-reality.

partially draw in this section—will represent the future of online interactions in the workplace and other settings and industries. Neal Stephenson (1992) first used the term metaverse in his novel about Hiro (protagonist) who delivers pizza while in the metaverse (a virtual reality system). Hiro is a warrior prince who attempts to neutralize a computer virus. However, the idea of people being able to perform actions in a "parallel world" that revolves around computer simulation can be found in texts from the 1980s such as Gibson's (1984) futuristic book *Neuromancer*. Gibson describes a dystopian reality where people can swing from their everyday life to the Matrix, a VR environment that can reward or punish individuals for their offline behaviors. Ball (2021) provides what I have found to be one of the most accurate definitions of the metaverse:

> A massively scaled and interoperable network of real-time rendered 3D virtual worlds which can be experienced synchronously and persistently by an effectively unlimited number of users with an individual sense of presence, and with continuity of data, such as identity, history, entitlements, objects, communications, and payments.

According to Ball's definition, the metaverse needs to be widespread (i.e., it is important that a critical mass of users access and use it). This condition applies to social media too. If not enough users access the metaverse, it becomes pointless because the metaverse is a social platform, which means that one underlying condition of its success relies on interactions between individuals. So far, the metaverse looks like a very lonely place as the first attempts to launch platforms fell short; even Meta's employees expressed concerns about the metaverse and its chances to gain market share.[8] Yet, it is undeniable that, in the future, we will witness online interactions no longer limited to Zoom-like 2D technologies because they are very limiting and prevent interactions that typically occur in person.

While so far I believe I have made it clear about the differences between synchronous collaborative systems such as Zoom and the metaverse (and virtual reality in general)—which revolve around immersiveness—from a data-harvesting standpoint, it is important to single out commonalities and differences between the metaverse and social media. Virtual reality platforms such as the metaverse and social media have a lot

[8] https://www.forbes.com/sites/paultassi/2022/10/07/report-even-metas-employees-dont-want-to-go-its-own-metaverse.

in common. For instance, they both need a critical mass of users to succeed (the reason why the metaverse is struggling). They both appeal to the general public, yet both are/will be used in the workplace. Enterprise social media platforms (cf. Leonardi et al., 2013) such as Yammer (now Viva Engage[9]) were built in a way that looked like Facebook to minimize disruption among users (organizational actors). Its creators developed Yammer to make user feel they were working in an environment similar to Facebook in order to encourage knowledge sharing and active discussion. However, enterprise social media did not succeed as knowledge management systems, and organizations soon replaced them with "lean" communication systems such as Slack[10] along with products fully integrated in operating systems such as Microsoft Teams.[11] In addition, social media and the metaverse offer similar value propositions for customers (or users) such as "connecting people" (or Meta's rebranded value proposition "bringing people close together[12]"). I would add that both social media and the metaverse share an underlying revenue stream based on ads. However, Meta and other organizations that create other virtual reality platforms seek to develop new business models based on renting their platforms to large companies and industries. For instance, Meta currently takes a nearly 50% cut on all sales from companies promoting their products and services on its platform called Horizon Worlds (Hamilton, 2022). Recent reports from Accenture[13] and Forbes[14] suggest that platform owners or otherwise large high-tech companies with knowledge and resources dominate the metaverse landscape, similar to how Microsoft dominates the productivity landscape with Microsoft Office 365 and Amazon dominates the cloud computing domain. To wit, while social media business models revolve almost exclusively on ads, the metaverse can count on revenue streams that go beyond ads and might be

[9] https://www.microsoft.com/en-us/microsoft-viva/engage.

[10] https://slack.com.

[11] https://www.microsoft.com/en-us/microsoft-teams/group-chat-software.

[12] https://about.fb.com/news/2018/01/news-feed-fyi-bringing-people-closer-together/.

[13] https://www.accenture.com/us-en/services/metaverse-index.

[14] https://www.forbes.com/sites/michellegreenwald/2022/02/03/why-nfts-and-the-metaverse-can-be-so-strategic-for-brands-even-providing-profitable-new-business-models-and-revenue-streams.

successful in penetrating the collaboration tool market in a more powerful way than social media.

Furthermore, it is worth noting that the metaverse offers data-harvesting possibilities that go above and beyond traditional enterprise systems and enterprise social media. People analytics (internal monitoring of employees) with the metaverse will afford capabilities to collect biometrics via sensors embedded in headsets and bodysuits. For instance, Meta can harvest features such as pupil movements, body poses, and nose scrunching. Via constant monitoring headset users' eye movements, AI can capture and analyze data about where people pay attention to the most and even what visual stimuli generate the greatest emotional feelings. Managers will be able to monitor, for example, whether virtual meetings attendees do something else (texting, answering emails) while online.

But all this leads to DEI implications concerning surveillance aspects that may be inappropriate and prone to discrimination (and, thus, exclusion). For instance, Kate Crawford (2021) has pointed to weaknesses of AI systems aimed at capturing people's behaviors through analysis of face expressions, motions, and so on through headsets. In addition, some technical issues concern AI's inability to operate with faces of females and people with darker skin as a seminal paper by Joy Buolamwini and Timnit Gebru highlights (Buolamwini & Gebru, 2018). Should metaverse-gathered analytics metrics be used to evaluate people's performance, they might be flawed, discriminate based on gender and race, and even penalize individuals with disabilities who are generally flagged as "unconventional" by automatic (i.e., algorithmic) systems (Sánchez-Monedero et al., 2020). Against these drawbacks, however, as biometric systems advance and improve in accuracy, people historically treated unequally with respect to salaries and job roles (women, non-White people) may be more objectively evaluated at the workplace. In fact, automatic performance evaluations brought on by the metaverse could reduce huMAN-led "un"conscious biases concerning rewards and promotions. This is one of those cases where algorithmic-based technologies can truly achieve less biased outcomes—the condition of this being that the current technologies improve from a diversity standpoint.

To that end, the metaverse's business models span from better targeted (and more invasive) advertising to selling tools to enable companies to increase surveillance over their employees (Murphy, 2022). While companies can readily adopt metaverse platforms for remote work (and surveillance), it is still not clear when (or, for some, whether) the metaverse will become an interoperable phenomenon. Interoperability is crucial for the metaverse to diffuse among the public at large for personal use and leisure (like social media). Interoperability means that metaverse users can cross platforms with a single avatar to make purchases and visit "places." Because online shopping is a key revenue stream for 3D virtual environments such as the metaverse (along with ads and data collection), it is important that interoperability is made possible. Avatars (i.e., users) should be able to move from one platform (i.e., Meta) to another (i.e., Microsoft) to visit stores (and make purchases) that different providers manage. Interoperability (or the lack thereof) has recently emerged as key in the fashion industry. For instance, in partnership with Meta, DressX6, an online digital fashion marketplace, raised US$4.2M in seed funding in 2020 to produce digital clothes that people can purchase and wear on the metaverse. However, as Daria Shapovalova (a co-founder of DressX) indicated in an interview with the *Wall Street Journal*[15]:

> Interoperability will be great as soon as we have it, but we're not there yet; as of now we need to prepare outfits for different environments. Various platforms have various rules—Meta is very different from Roblox.

Going back to the metaverse versus social media comparison, while the granularity of metaverse data makes the platform a perfect candidate for surveillance (due to sensors and access to biometrics data), it is difficult to think how all these data can be stored and retained as digital traces. Social media is mainly an asynchronous system where people drop posts, messages, and so on, which other people see at a later time. In the metaverse, everything happens synchronously, which does not make it possible to collect digital traces for two reasons. First, there is simply too much data to collect and classify for future extraction (metaverse data is 3D video data). Second, in the metaverse, it is possible to create private chatrooms, which are based on user-generated content, and it is impossible to access all these chatrooms and store conversations and interactions. In addition, metaverse data includes videos because people (or their avatars) do not only talk but they also perform actions. The fact that it is hard (or impossible) in the metaverse to capture and store data in the form of

[15] https://www.wsj.com/articles/fashion-in-the-metaverse-what-avatars-are-wearing-11669995878.

digital traces leads to problems associated with its moderation. The metaverse has already been labeled as such a place where harassment takes place.[16] How is it possible to prevent harassment from happening if it is nearly impossible to collect records about what happens on these platforms? Social media is a place where harassment, cyberbullying, and other such practices occur as well. And the lack of moderating power has made news.[17] But harassment on social media has a totally different impact on people's emotion than that in the metaverse. The immersive nature of the metaverse offers quasi-real experiences (e.g., visiting a city using immersive virtual reality is different than reading a book or watching a movie). But the same feelings of immersiveness can hurt harassed people more than if harassment on a social media platform because, in the metaverse, everything looks more real.

In healthcare, the metaverse has an important role, specifically because of the sensors devices that are used to access these platforms. As I mentioned earlier, telehealth expanded substantially (for emergency reasons) during the pandemic but, similarly to primary and secondary education, in-person visits resumed in its aftermath. While the metaverse can offer more accurate visits via sensors, its widespread diffusion might discriminate against people who cannot afford to access the metaverse. For instance, patients would need to wear devices (e.g., bodysuits) to allow clinicians to capture vital information. These devices are expensive and, for a long time, won't be available for everyone. For instance, it is foreseeable that, in the US (a private/public healthcare system where private health is prevalent and very expensive), only people with health insurance (private) will be able to access immersive remote care. In addition, while telehealth is supposed to reach populations in rural areas and minimize travel for the elderly and people with limited mobility, remote immersive care requires broadband, which might be unavailable in these areas.

Metaverse applications to higher education are prone to exclusion practices, similar to the ones present in healthcare settings, with the potential for discrimination against financially weak students, students living in rural areas, students with disabilities, and so on. If we shift the focus from

[16] https://www.theverge.com/2021/12/9/22825139/meta-horizon-worlds-access-open-metaverse.

[17] https://www.nytimes.com/2023/02/14/technology/disinformation-moderation-social-media.html.

students to professors, along with issues previously mentioned concerning remote work, academics are and will face challenges to attend conferences in hybrid mode. I discuss DEI implications of hybrid conferences in Chapter 2 where I refer to my study with Sam Zaza, Silvia Masiero, Lydia Li, and Kathy Chudoba (Marabelli et al., 2023) to showcase how important it is to listen to the minority, often people from the Global South, who advocate for hybrid rather than in-person conference. Here, I use again this study to point to a different thing that relates more to virtualized conferences and to whether they will be the future of academic gatherings. With respect to hybrid conferences, on the one hand, the need for expensive equipment to attend remotely using virtual reality may exclude people from financially weak universities, something that will make it pointless to use hybrid techniques to include participants from the Global South—even if might allow other people otherwise not able to travel, such as pregnant women, to participate.

On the other hand, hybrid conferences might lead to first and second class citizens with the former enjoying social events with food and drinks after paper sessions while the latter simply logging off. In this way, hybrid conferences pursue diversity (e.g., allowing participation from more people from the Global South), while being detrimental to inclusion. Here, it is important to reflect on whether the diversity and inclusion constructs should be decoupled in certain settings. In fact, what's the point of achieving diversity if it leads to segregation? The 1954 landmark US Supreme Court case Brown v. Board of Education of Topeka reversed centuries of segregation practices in the US by establishing that racial segregation in public schools is unconstitutional even if the segregated schools are otherwise equal in quality. By reversing Plessy v. Ferguson (1896), which supported the "separate but equal" doctrine, the Brown ruling highlighted that "separate educational facilities are inherently unequal." In the US, unequal treatment of citizens based on race violates the Fourteenth Amendment, which addresses citizens' rights and equal protection under the law and was proposed in response to issues related to formerly enslaved Americans following the American Civil War. And because White schools had better quality (more qualified teachers, better infrastructures, segregation by race would tread Black Americans differently and lead to lower education standards). This is precisely what might happen if the metaverse is used as a remote solution for attendees from the Global South. Let alone the costs associated with accessing conferences using virtual reality, we need to make sure that metaverse attendants do not become second class citizens that organizers bring in simply to increase diversity at the expense of inclusions in social gatherings.

Here, I don't have an answer to whether the metaverse can help make remote attendance to conferences a "better" experience for remote attendees. Certainly, the risk of segregation is high. Yet, one may argue that attending conferences remotely, especially using virtual reality, is "better than nothing"—the reason why I strongly advocate for hybrid conferences whenever possible. I came to this conclusion after carefully examining research on hybrid conferences, the extent to which people from the Global South but also people from financially weak universities in the Global North can (or cannot) participate in these gatherings, and how things had changed during the pandemic. From my study on hybrid conferences, I learned about a multitude of people who, during the pandemic, could finally attend international conferences for the first time—conferences that had always been held only in person. One issue that might emerge with the use of virtual reality to attend conferences is broadband, which is required to sufficiently project remote environments in 3D. From my study, having interviewed several academics from the Global South, the issue of lack of broadband in their countries was paramount even with Zoom. One study participant highlighted: "Internet connection issues and challenges with [a] high risk of being disconnected in the middle of a presentation." The metaverse could be one of these technologies still immature for the time being if used to bridge Global North and Global South with respect to academic gathering due to the lack of global infrastructures supporting the volume of data exchange needed to enable 3D systems.

An interesting aspect of the development of the metaverse (and other virtual reality platforms) concerns the role AI will play. AI is now ubiquitous and it is hard to think that it won't find considerable use in the metaverse, but how? The most obvious use of AI in the metaverse involves generating visual surroundings, such as virtual environments. With generative AI being able to generate images from prompts, something already offered by companies such as DeepAI,[18] the metaverse will employ AI to support user-generated content. Different from social media, this user-generated content will concern landscaping, objects, and more, and it will

[18] https://deepai.org/machine-learning-model/text2img.

be sufficient for users to "speak" to an AI to convert one's imagination into a scenario. AI can also be used in the metaverse to leverage real-time capabilities to capture and interpret biometric data collected by users (e.g., in people analytics contexts). In a recent paper with Pamela Lirio, Associate Professor at the School of Industrial Relations at the University of Montréal, I explore how companies use AI in metaverse platforms to allow remote work and monitor employee performance (Marabelli & Lirio, 2024). We focused on the potential DEI implications of the metaverse in the workplace with a specific interest in exploring how AI may contribute to discrimination and inequalities. On paper, AI in the metaverse can allow objective performance management. As I point out earlier in this chapter, the metaverse in the workplace can capture very intimate data on users such as eye movements, which AI can process as an indicator of people's focus during online meetings. Management's assessment would, therefore, be less biased. If the data says that a woman performs better, managers and executives with prejudices against promoting female workers will have to follow the data. The same applies to non-White people or people usually face discrimination in the workplace.

One important difference between how AI is commonly used and how it is used in the metaverse is that, in the metaverse, AI does not rely only on historical data to make judgments. In the metaverse, AI uses real-time data, so outcomes are not only affected by past, potentially discriminatory practices. For instance, in hiring contexts, AI used to scan resumes for hiring people can discriminate against women and Black people due to historical hiring data (Raub, 2018). In the metaverse, interview outcomes can be informed by more "objective" sensor data, and managers will not be led to make gender or race-based hiring decisions because the underlying algorithms would not present as inherently biased. The same applies to workforce evaluations. In addition, collecting biometrics on employees may lead organizations to closely monitor important health parameters. Employee burnout can lead to mental health issues and concerns for employers. AI can monitor employee stress levels as an intervention tool and provide counseling. While mental health issues are "neutral" across populations, underrepresented populations are more prone to not receive proper care because of the general lack of trust that these populations have in healthcare structures, especially in the US. With the metaverse, companies have a chance to fill this gap by assisting their employees and providing equal access to care for their entire workforce.

The real-time nature of the metaverse, however, poses DEI issues concerning the impossibility to record everything that happens when employees interact, which raises questions about whether virtual platforms might encourage harassment in the workplace—according to a study conducted by the International Labor Organization, 20% of the worldwide workforce lamented that they experienced harassment during work hours.[19] What is more, health-related data that employers collect via headsets and bodysuits constitute sensitive data (at least in the US). Would companies collecting this data be subject to the Health Insurance Portability and Accountability Act of 1996 (HIPAA) standards concerning data protection? This issue is often mentioned when referring to companies such as Apple (for the Apple Watch) and Google (who owns Fitbit). With the metaverse, it looks like HIPAA standards should apply to any employers that collect health data. Solutions concerning the ethical design of these platforms (e.g., with respect to privacy) are needed (see the example of the design of the Apple Watch and Fitbit in Chapter 2). This urges me to discuss an extremely timely issue on how technologies are currently designed, one example being generative AI.

The Role of Design in the Development of Emerging Technologies

The need for ethical technology designs is a longstanding subject that scholars and practitioners have broadly discussed (cf. Martin, 2019). One main issue concerning (ab)use of technologies concerns how companies can turn innocuous innovations into surveillance tools or even how users enact technologies in ways that they are not designed for. With respect to companies and sticking with the metaverse, this platform is built with analytics capabilities for safety reasons. It is important to monitor what people do in the metaverse to prevent online harassment and bullying along with other behaviors that can hurt people. Yet, companies can use these very tools to mine data about their employees and generate performance management systems that penalize women or differently able people. With this, I am not saying that companies do this purposely. Yet, the hypercompetitive environment in which we live nowadays along with

[19] https://www.ilo.org/global/about-the-ilo/newsroom/news/WCMS_863177/lang--en/index.htm.

companies' renowned revenue focus represent great incentives to pursue efficiency (automated performance assessment is less time-consuming than manual assessment and needs less human capital) even if it happens at the expense of some people. With respect to individuals and still sticking with the metaverse, the platform is designed to make interactions more real. But what if this feature is exploited by people to threaten others, make inappropriate comments, and so on, which would hurt people more than hate speech on social media precisely because platforms such as the metaverse are immersive and feelings of being in the "real world" are part of their design.

The literature on affordances (Gibson, 1977), which researchers have applied to different technologies in the last few decades beginning with Norman's (1991), explains how designing artifacts does not determine their use. A typical example is that of a chair that is designed so that people can sit on it but that can be used to break a window and escape a room on fire. This is one way to use an artifact in a way that diverges from the use the designer had in mind. This example of the chair showcases that uses deviating from the designer's purposes are not always negative—some uses led to innovations. For instance, when first cell phones became widespread (in the late 1990s), people used to use them at night to see in the dark when walking to the bathroom. Years later, more modern phones embedded flashlights. After the first mobile phones made the time visible on the lock screen, people started using cell phones as clocks, and the sales of wrist watches dropped.

With the advent of AI (and, in particular, generative AI), we are witnessing widespread misuses of large-scale available software. Examples include deepfakes (including revenge porn videos and videos including minors), fake news (which has an important impact on political elections), and essays written with ChatGPT. Is all this happening because novel technologies were made available to the public before vetting their potential for being used in ways they were not designed? This is certainly a valid point. As I've previously explained, US capitalism includes ideas such as that of innovation becoming compromised by too much vetting with respect to societal issues. Most high-tech companies believe that "remedies" should be put in place once problems with technology uses emerge. This is quite at odds with an approach involving ethical design—design that is aimed to prevent technology misuse and that requires comprehensive tests before a technology is released to the general public.

But what if companies release technologies to the public for reasons spanning from short-term profits to achieving first-mover leadership and even to purposely exclude people from accessing services while knowing that people can easily use them in ways to hurt others? Safiya Noble (2018) makes this point in providing hard evidence of the "politics of technology design" (p. 92) where high-tech companies such as Google are careless when conceiving algorithms that need to serve the "majority" (White users of their search engine platform) and purposely neglect minorities. Other examples can be found in other technologies such as Amazon Ring, which enables people to surveil and that implicitly enables domination with people (Amazon Ring owners) allowed to share recordings with law enforcement.[20] Companies that design technologies aimed at surveillance and make them available to the public implicitly enable people in power to exercise this power over others. The surveillance issue is particularly important in the US, where laws regulating data brokering are nearly absent. To some extent, these technologies are designed to allow or enable discriminations that originate from private citizens (people who purchase and use Amazon Ring) but then spread to local institutions such as law enforcement who use the data from the technologies to target minorities. For instance, a 2021 report released by New York University's School of Law[21] indicates that Amazon Ring could lead to overpolicing, racial disparities and bias, lack of accountability, and transparency. Amazon Ring exemplifies a technology designed to empower certain individuals (i.e., people who can afford these surveillance systems) at the expense of others. In addition, I argue that whoever wants to own a surveillance system is already in power because they have assets to protect. This does not mean that people don't have the right to protect their private property (something spelled out in US's Fifth Amendment). At the same time, the Fifth Amendment should not be exercised to target and oppress others.

Other examples of technologies designed in ways that discriminate against people include some deployed (or that became widespread) during the pandemic. For some, the racist intent is more evident than others. For instance, oximeters (see Chapter 1) that are more reliable if used by

[20] https://arstechnica.com/tech-policy/2022/07/amazon-finally-admits-giving-cops-ring-doorbell-data-without-user-consent/.

[21] https://www.policingproject.org/ring.

Whites follow past trends of exclusion of Black communities from clinical trials. This poses two issues. First, oximeters are designed for whoever can afford them, and financial status is often a proxy to race. Second, oximeters, which measure the level of oxygen in the blood, a key leading indicator to determine COVID-19 severity, are helpful to make a decision on whether to head to the emergency room when the oxygen in the blood drops below a certain level (80%). Because oximeters work better with Whites and Whites are more likely to be able to afford to purchase one, Whites—still in 2023—are better served by the healthcare system. Statistics suggesting that Black Americans are more prone to severe COVID-19 cases should account for disparities in access to care rather than relying on claims concerning biological differences with respect to skin color.

Design has always played a relevant role in how technologies affect DEI, sometimes in ways that purposely penalize underrepresented populations. Over 40 years ago, Langdon Winner's (1980) seminal paper on the power of artifact provided a meaningful example on how public infrastructures could be designed to exclude people already at the margin:

> Anyone who has traveled the highways of America and has become used to the normal height of overpasses may well find something a little odd about some of the bridges over the parkways on Long Island, New York. Many of the overpasses are extraordinarily low, having as little as nine feet of clearance at the curb ... Even those who happened to notice this structural peculiarity would not be inclined to attach any special meaning to it. ... It turns out, however, that the two hundred or so low-hanging overpasses on Long Island were deliberately designed to achieve a particular social effect. Robert Moses, the master builder of roads, parks, bridges, and other public works from the 1920 s to the 1970s in New York, had these overpasses built to specifications that would discourage the presence of buses on his parkways. Automobile-owning whites of "upper" and "comfortable middle" classes...would be free to use the parkways for recreation and commuting. Poor people and blacks, who normally used public transit, were kept off the roads because the twelve-foot tall buses could not get through the overpasses (Winner, 1980, pp. 123–124).

The quote from Winner speaks for itself. People in power have the power to discriminate by creating infrastructures that penalize some. Having recognized the political role of technology design, the design of the metaverse becomes a key issue, especially due to its algorithms that, for example, have the potential to evaluate employee performance (people analytics). But designing the metaverse does not concern just algorithms. The design of the devices to access the platform plays an important role. For instance, because headsets have been traditionally sold to (generally males) gamers, the lens spacing was not designed for the anthropometrics of women's faces, which led females to be more prone to experience discomfort and nausea (Stanney et al., 2020). Figure 6.1 is an example of headset. It is no coincidence that is worn by a (White) male.

Examples of design purposely made to discriminate or to make money out of powerless people proliferate. Along with the case of Instagram hiding research about how its algorithms negatively affected teenagers with eating disorders—they were designed to over-engage users (see Chapter 5), a recent lawsuit filed by Massachusetts includes allegations indicating that Mark Zuckerberg, Meta's CEO, willingly dismissed recommendations from senior team leaders to dismiss recommendation algorithms that potentially harmed the mental health of women. Executives, including Instagram head Adam Mosseri, pushed for the company to ban filters that mimic plastic surgery due to concerns that they could

Fig. 6.1 Headsets to access the metaverse[22]

[22] Credits: pixaby.com.

harm the mental health of women and teens. Karina Newton, Instagram's Head of Policy at the time, reported that:

> It's been our strong recommendation from comms, marketing, policy, and engagement with nearly 20 outside experts and academics that we pass this policy … We're talking about actively encouraging young girls into body dysmorphia.[23]

The lawsuit says that Zuckerberg nixed the effort: he canceled a meeting in which staffers were set to argue for the proposal based on the claim that he had seen "no data" suggesting that the filters were harmful.

Here, the difference between Winner's reporting of bridges designed to prevent Black Americans to access some areas of New York City and social media companies allowing algorithms to profit at the expense of people at the margin is thin. In the 1940s, willingly segregating populations was a de jure practice as, before the Brown vs. Board of Education landmark ruling, federal and local US administrations refused to acknowledge the Fourteenth Amendment. Nowadays, it is impossible to create (digital) infrastructures that purposely discriminate (at least de jure). Yet, deploying poorly vetted and irresponsibly designed algorithms and keeping them in production while being aware they harm weak populations can be equated with the Moses bridges. Moses wasn't particularly against Black communities. He saw profits in segregating Black communities. The same happens today when algorithms are "let go" on social media platforms, never mind if they target young women with eating disorders or with poor self-confidence as long as ads from companies offering surgical "remedies" to their issue will contribute to social media revenue streams.

As Langdon Winner reports, in the past, politicians could segregate by race using infrastructure. Note that segregation in most cases is not neutral: the Plessy v. Ferguson *separate but equal* doctrine has always advantaged Whites. Nowadays equivalent of bridges with law clearance in the mid of the nineteen hundred century is investment in infrastructures rewarding rich people with SUVs, such as plans to build or redo highways with multiple lanes crossing large cities and allowing heavy traffic rather than investing in public transportation. These investments

[23] https://www.wsj.com/tech/metas-deception-about-instagrams-harmful-qualities-alleged-in-massachusetts-suit-e4729acc.

purposely focus on gentrifying urban areas with the ultimate goal to make the life of people using public transportation hard, which would lead to ghetto neighborhoods at a walking distance from low-income jobs. American cities are designed for cars, which, as Winner highlights, penalizes the poor who need to use public transportation (proxy for minorities in the past and nowadays). This process began in the 1920s with Ford's Model-T. A good example of a car-centric urban area is Los Angeles where long distances can be covered in a moderate timeframe only with cars. Boston has highways that cross the city. Post-pandemic habits include working from home at least some days a week. This has encouraged people who can work remotely to move to suburban areas, which further encouraged them to purchase and use cars while reducing the demand and revenues for public transit.

What policies and strategies will lead the development of metaverse infrastructures? Right now, the prices of devices allowing access to virtual reality are rising. For instance, in 2022, Meta Oculus 2 could be purchased for US$250. Newer models with additional features saw their price rise to US$450. Meta Quest 3 has spanned from US$500 to US$1,000 (the latter in the "pro" version). All this means that the more money one has, the better one will be able to access Meta platforms. Apple's mixed reality device (Apple Vision Pro) costs US$3,500—something that most people might not be able to afford.

Circling back to Winner's point, it's not hard to imagine a two-tier system where large metaverse platforms will make sure that some (rich, White) people will access the metaverse in ways that are unaffordable by other populations. In higher education, community colleges students will be able to experience the metaverse differently than students attending prestigious universities. This will affect their chances to effectively networking with peers and professors and, thereby, shrinking their chances to access well-paid jobs. Is all this carefully planned by companies wanting to make sure to perpetuate segregation to please White majorities? Whites are generally afraid to lose their privileges by being mixed with people who have historically been excluded. While most innovations happened overnight with lawmakers unprepared to prevent discriminations, as the metaverse remains *in the making*, we have a chance to prevent such discrimination from happening by making sure that its infrastructures do not exclude certain people.

Other Emerging Technologies

The metaverse constitutes just one among many technologies that emerged since 2020 (during the pandemic), which harmed most industries but gave a boost to high-tech companies due to the virtualization of work, schooling, healthcare, and so on. For instance, one year into the pandemic, Amazon reported US$108.5B in sales in the first quarter of 2021 and US$8.1B in profit, a 220% increase from the same period in 2020.[24] Zoom reported a profit of US$882.5M million in the last quarter of 2020, an increase of US370% compared to the same period in 2019. Netflix, Shopify, and PayPal are among other companies that reported substantial growth in revenue and profits during the pandemic. In times of crisis and economic downturns, there are always companies that benefit. For instance, during the 2008 Great Recession, discount stores such as Dollar Tree and Walmart increased their revenue because customers switched from fancy stores to more affordable ones. But the pandemic gave unprecedented power to high-tech companies. The ubiquitous nature of the products and services that these companies deliver placed them in a position to collect data and profile current and perspective customers in (at best) questionable ways as I explain throughout this book.

Interestingly, while the status quo generally regains prominence when economies overcome recessions, the pandemic set new standards for instance concerning remote work; WeWork[25] a New York City company providing office space for businesses once valued near US$500B, filed for Chapter 11 bankruptcy protection on November 7, 2023. In the same year, Nvidia obtained record revenue of US$13.51B, up 88% from the first quarter and up 101% from the same period in 2022.[26] Clearly, the world has changed. Virtualization is prominent and prevalent. Climate changes encourage people to commute less and work from home. As the psychological consequences of workplace socialization happening remain virtually unclear, businesses worldwide have reconfigured their work practices for good. But the pervasive and invasive consequences

[24] https://www.businesswire.com/news/home/20210429006037/en/Amazon.com-Announces-First-Quarter-Results.

[25] https://www.wework.com.

[26] https://nvidianews.nvidia.com/news/nvidia-announces-financial-results-for-second-quarter-fiscal-2024.

of technologies in the workplace in the aftermath of the pandemic has had relevant spillover effects to other technologies and industries. In my opinion, it is the substantial amount of money poured into high-tech companies that explains, at least in part, the huge development of technologies such as generative AI and advanced algorithms affecting people's everyday life.

Many people now use ChatGPT: students use it to summarize reading assignments, innovators use it to seek ideas, and regular people use it to look for answers to everyday problems. But ChatGPT is highly biased for two reasons. First, it relies mostly on Web knowledge. But the Web is not a neutral nor a democratic place (see Chapter 2 and the seminal paper by Bender and Friedman (2018)). Second, ChatGPT is constantly trained by people who ask questions via prompts. This means that the answers ChatGPT provides are biased by its heavy users, who provide the most training data. In the long term, ChatGPT might lead to a global filter bubble via addressing questions with principles that rely on Western cultures, White dominancy, and so on "in mind." The problem is that ChatGPT is currently used in schools, which often have students from diverse backgrounds who would have to rely on software that reflects cultures and beliefs highly biased toward a capitalistic culture they might not belong to. Until these students will become heavy users, thereby being able to influence ChatGPT training models, they will be passive recipients of dominant cultures they might not belong in.

Other industries where algorithms dominate include the music business. While digital music (e.g., CD, Blue-ray) has been around for decades, right now algorithmic-based platforms such as Spotify rule. Spotify[27], a Swedish audio and media streaming platform founded in 2006 by Daniel Ek and Martin Lorentzon, has over 500 million monthly active users and 226 million paying subscribers (2023 data). Among other platforms, Spotify is used to rank top billboard songs, which are in turn broadcasted by radio and TV stations. Spotify's algorithm recommendation offers users songs that are in line with one's preferences and analyzes lyrics, rhythm, genre, and so on; in this way, it might represent a filter bubble. But Spotify's algorithm does more. The first 30 seconds of a track matters to Spotify more than anything else. If a user gets past the 30 second mark of a song, that's a positive bit of data and represents the

[27] https://open.spotify.com.

point at which a stream is monetized. In turn, singers who want their songs promoted by Spotify need to avoid long instrumental intros (think about La Vie en Rose sang by Grace Jones in 1977, a worldwide hit with a 43 seconds intro[28]—it would never be a top hit on Spotify nowadays), which would risk the listener to drop the song too soon and, thereby, not allow Spotify to rank it. The length of a song matters too. Mark Ronson, a British musician, DJ, record producer, songwriter, and remixer, told *The Guardian* that:

> All your songs have to be under three minutes and 15 seconds because if people don't listen to them all the way to the end they go into this ratio of 'non-complete heard', which sends your Spotify rating down.[29]

In fact, a study conducted by Ostereo, a digital label based in the UK, found that pop songs are one minute and 13 seconds shorter on average than they were 20 years ago.[30] This means that art is currently compromised by algorithms that serve music labels and influence artists and the way they conceive songs. If you want to be in the game of pop music, you need to obey to algorithmic-based rules designed to harvest profits. A related, yet different phenomenon is TikTok, that can literally "launch" new singers and songs if short videos become viral. Kenya Grace, a British singer, composed and recorded her hit "Strangers" in her bedroom and, soon after its appearance on TikTok it hit the top of the UK singles chart, in October 2023. Here, TikTok recommendation algorithm generated revenue streams for Kenya Grace, and for the TikTok itself because once a video goes viral, both the video and the platform benefit from it.

Lastly, Humane's AI *Pin*[31] represents cutting-edge, generative AI-powered devices that might dominate the future of pervasive AI. It is basically a wearable smartphone without a screen[32] that can act with voice commands and record one's surroundings with a camera. One can wear the device on shirts and jackets. Besides outstanding privacy issues with respect to recording in public places (something allowed in some

[28] https://www.youtube.com/watch?v=YEM8TspcCBY.

[29] https://www.theguardian.com/music/2019/feb/22/mark-ronson-I-was-floundering-I-was-drinking-too-much-and-giving-orders.

[30] https://www.prsformusic.com/m-magazine/features/song-length-the-spotify-effect.

[31] https://hu.ma.ne/aipin.

[32] https://www.theverge.com/2023/11/8/23953022/humane-ai-pin-price-specs-leak.

countries but forbidden in others), these types of surveillance devices are designed to further empower the powerful similar to the music industry where gatekeepers (Spotify) set the rules of the game. In fact, in the music industry, consolidated bands can break the rules and record songs that don't comply to Spotify rules, while emerging bands need to follow suit.

Emerging technologies mirror past societal inequalities. This is not surprising as technologies reflect who we are and how we aim to use technologies to either exercise power or do some good for society. Technologies, even centuries ago, reflected power dynamics. The first industrial revolution that occurred across the UK in the eighteenth and nineteenth centuries led to increased efficiency in the production of goods. Yet, as Alice Clark (1919) noted, with the first industrial revolution, women lost their status within their families because of their decreased economic relevance. Before the first industrial revolution, women played a key role in many aspects of industry and agriculture, but, once the first manufacturing plants opened, men were the main resources for these growing companies, and labor divisions increased. Men were well-paid and women were reduced to unpaid household work. The fact that men were more prone to take on jobs requiring physical effort led to a social system (and mindset) by which men made money and women did not. This became a stereotype throughout the years and the centuries and resonates as the status quo today. The biological fact that women give birth adds to all this: women are then supposed to care for infants by nursing and by being the center of children's education while men work and support the family. This patriarchal vision is still present nowadays in Western cultures, especially in low-income families, where men do manual and physically demanding jobs that women often cannot do as well as men. The problem is that this culture propagates to other jobs that women *can* do well (intellectual work), that they can do when working from home, and that do not prevent them, for example, from breastfeeding their children if they want to.

Algorithms pick up on these and other historical injustices and help perpetuate them in our society. Women struggling to get into STEM programs because of past events such as the division of labor during the first industrial revolution is not different than the power exercises by dominant labels in the music industry, which rely on algorithms to obtain music sales, which disrespects the art of composing and singing a song.

Like manufacturing companies that ruled in the late 1700s, the music industry uses AI to rule nowadays and perpetuates power asymmetries between business and people/employees.

Conclusion

In this (last) chapter, I describe some perspective innovations in AI and algorithms realm, one prominent being the metaverse but, more generally, all technologies that focus on enhancing virtuality. After the COVID-19 pandemic, the world has realized that most work (and leisure) activities can be efficiently conducted online and that people can use novel ways to interact with one another and the world (e.g., virtual, augmented, and mixed reality). But at what price?

References

Bender, E. M., & Friedman, B. (2018). Data Statements for Natural Language Processing: Toward Mitigating System Bias and Enabling Better Science. *Transactions of the Association for Computational Linguistics, 6*, 587–604.

Bestsennyy, O., Gilbert, G., Harris, A., & Rost, J. (2021). *Telehealth: A Quarter-Trillion-Dollar Post-Covid-19 Reality?* McKinse Report, Available at https://www.mckinsey.com/industries/healthcare-systems-and-services/our-insights/telehealth-a-quarter-trillion-dollar-post-covid-19-reality

Bridges, K. (2011). *Reproducing Race: An Ethnography of Pregnancy as a Site of Racialization*. University of California Press.

Brucks, M. S., & Levav, J. (2022). Virtual Communication Curbs Creative Idea Generation. *Nature*, Published online on April 27, 2022, Available at https://www.nature.com/articles/s41586-022-04643-y.pdf

Buolamwini, J., & Gebru, T. (2018). Gender Shades: Intersectional Accuracy Disparities in Commercial Gender Classification. *Conference on Fairness, Accountability and Transparency*: Proceedings of Machine Learning Research, pp. 1–15.

Clark, A. (1919). *Working Life of Women in the Seventeenth Century*. Economic History (Routledge).

Crawford, K. (2021). *The Atlas of AI: Power, Politics, and the Planetary Costs of Artificial Intelligence*. Yale University Press.

Gibson, J. J. (1977). *The Theory of Affordances*. Lawrence Erlbaum.

Gibson, W. (1984). *Neuromancer*. New York, Ace.

Hamilton, I. (2022). *Meta Will Charge Creators Fees of up to 47.5% to Sell Virtual Wares in Its Metaverse*. Business Insider, available at https://www.businessinsider.com/meta-metaverse-charges-creators-47-percent-sales-fee-cut-zuckerberg-2022-4, Accessed January 22, 2023.

Labianca, G., & Brass, D. J. (2006). Exploring the Social Ledger: Negative Relationships and Negative Asymmetry in Social Networks in Organizations. *Academy of Management Review, 31*(3), 596–614.

Leonardi, P. M., Huysman, M., & Steinfield, C. (2013). Enterprise Social Media: Definition, History, and Prospects for the Study of Social Technologies in Organizations. *Journal of Computer-Mediated Communication, 19*(1), 1–19.

Lin, Y., Frey, C. B., & Wu, L. (2023). Remote Collaboration Fuses Fewer Breakthrough Ideas. *Nature, 623*(7989), 987–991.

Marabelli, M., & Lirio, P. (2024). AI and the Metaverse in the Workplace: DEI Opportunities and Challenges. *Personnel Review*, Available online, https://www.researchgate.net/publication/374911261_AI_and_the_Metaverse_in_The_Workplace_DEI_Opportunities_and_Challenges

Marabelli, M., & Newell, S. (2023). Responsibly Strategizing with the Metaverse: Business Implications and DEI Opportunities and Challenges. *The Journal of Strategic Information Systems, 32*(2), 1–16.

Marabelli, M., Vaast, E., & Li, L. (2021). Preventing Digital Scars of Covid-19. *European Journal of Information Systems, 30*(2), 176–192.

Marabelli, M., Zaza, S., Masiero, S., Li, J., & Chudoba, K. (2023). Diversity, Equity, and Inclusion in the AIS: Challenges and Opportunities of Remote Conferences. *Information Systems Journal, 33*(6), 1370–1395.

Marr, B. (2022). *The Amazing Possibilities of Healthcare in the Metaverse*. Forbes, available at https://www.forbes.com/sites/bernardmarr/2022/02/23/the-amazing-possibilities-of-healthcare-in-the-metaverse/?sh=daeebe99e5c2

Martin, K. (2019). Designing Ethical Algorithms. *MIS Quarterly Executive, 18*(2), 129–142.

Murphy, H. (2022). Facebook Patents Reveal How It Intends to Cash in on Metaverse. *Financial Times*, available at https://www.ft.com/content/76d40aac-034e-4e0b-95eb-c5d34146f647

Noble, S. U. (2018). Algorithms of Oppression. In *Algorithms of Oppression*. New York University Press.

Norman, D. A. (1991). Cognitive Artifacts. In J. M. Carroll (Ed.), *Cambridge Series on Human-Computer Interaction, No. 4. Designing Interaction: Psychology at the Human-Computer Interface* (pp. 17–38). Cambridge University Press.

Raub, M. (2018). Bots, Bias and Big Data: Artificial Intelligence, Algorithmic Bias and Disparate Impact Liability in Hiring Practices. *Arkansas Law Review, 71*(2), 529–570.

Sánchez-Monedero, J., et al. (2020). What does it mean to'solve'the problem of discrimination in hiring? social, technical and legal perspectives from the UK on automated hiring systems. *Proceedings of the 2020 Conference on Fairness, Accountability, and Transparency*.

Stanney, K., Fidopiastis, C., & Foster, L. (2020). Virtual Reality Is Sexist: But It Does Not Have to Be. *Frontiers in Robotics and AI, 7*(4), 1–19.

Stephenson, N. (1992). *Snow Crash: A Novel*. Bantan Spectra.

Winner, L. (1980). Do Artifacts Have Politics? *Dedalus, 109*(1), 121–136.

CODA

I wrote this book to generate awareness about the potential threats that AI and algorithms pose to society. While I acknowledge the infinite possibilities for automated systems to make our lives better, the future remains uncertain. In America (but also elsewhere), businesses face various challenges, such as hypercompetitive environments, high inflation, high interest rates, and relates issues. For instance, many banks must contend with long-term mortgages and investments in long-term government bonds that have become unsellable (cf. the Silicon Valley Bank example in Chapter 5). Despite these challenges, the US economy seems to have avoided a recession—at least according to some sources. However, for markets, businesses, and customers, perception constitutes reality. Even if the US currently experiences unprecedented a high employment rate, many commonly feel the country lacks stability. As for why, factors include the upcoming presidential election in 2024 and "post-pandemic syndrome," which left psychological scars in many people, changed how (and where) people work, and resulted in decreased offshoring operations (due to supply chain disruptions). Indeed, we can reasonably expect this latter factor to increase the cost to produce technologies (among other products) and, thus, impact both the Global North and the Global South. In addition, two wars have displaced people and raised international tensions among big powers such as the US, China, and Russia.

M. Marabelli, *AI, Ethics, and Discrimination in Business*, Palgrave Studies in Equity, Diversity, Inclusion, and Indigenization in Business,
https://doi.org/10.1007/978-3-031-53919-0

AI and algorithms continue to become increasingly effective shortcuts to pursue efficiency such that task workers, when organizations do not fire them, face lower wages. In fact, along with poorly testing and deploying algorithms in inconsiderate ways, the race for automation penalizes poorly educated individuals who cannot move from routine jobs (that face automation) to more knowledge-based ones. As I attempted to explain in this book, all these dynamics and changes do not affect all individuals equally. Poorly tested algorithms penalize people at the margin and, thereby, perpetuate longstanding social injustices. Automation follows suit. Critical race theory (CRT) serves as a tool to explain how these social injustices stem from our past and how people and organizations continue to reproduce them in various ways (e.g., via laws and regulations that White males create). Indeed, we have no reason to believe that our Founding Fathers considered Black people when writing the US Constitution since, at that time, Black people largely lived in slavery conditions. The fact that many current institutions continue to feature designs that advantage Whites explains why algorithms today continue to reflect the past. Designing colorblind algorithms is not enough: they would unfairly penalize some people more than others just as state taxes on basic products such as eggs, milk, and bread do.

One main point that I make in the book involves government intervention. For instance, the US lags behind the European Union greatly in taking action on data privacy and AI systems, which partially explains why so many people in America live in misery (both financially and psychologically). In the US, algorithms determine whether one can access credit, healthcare, jobs, and so on. Discriminating algorithms reproduce past injustices in a very systematic way. Furthermore, in addition to AI, other new technologies such as the metaverse and blockchain have emerged that may negatively affect people's everyday life. While social media took everyone by surprise (hence, the lack of regulations), we still have time to implement rules for emerging technologies. If unregulated, generative AI poses important privacy issues, and, if used to make decisions about people, it will be prone to discrimination. If unregulated, the metaverse poses important diversity, equity, and inclusion (DEI) issues concerning accessibility and widespread efforts to collect biometric data. Blockchain systems failed to deliver on their premises (e.g., with respect to cryptocurrencies and barriers to access trading). On the contrary, in the US, many failed companies, such as FTX, have overly penalized Black Americans who sought easier access to financial trading, but lost all their money.

In fact, regulations concerning investments in cryptocurrencies are still largely absent and, as a consequence, investors are not protected. Now, we have a chance to regulate all these algorithmic-based technologies before they cause more damage. But high-tech companies do not want regulations. On the one hand, they identify themselves as very liberal and inclusive. On the other hand, they still support White and male culture (e.g., the OpenAI board) and a laissez faire approach to business. I believe such cases represent one of the biggest contradictions of America.

I found it difficult to write this book in a way that reflects the "current" day due to the pace at which new technologies, regulations, political, and macro-economic dynamics emerge. However, I have tried to report recent news on the topics I discuss. For instance, as I finalized one chapter where I lament social media companies' poor transparency with respect to data sharing, Meta came up with an API that allowed "selected" researchers to access real-time feeds on Facebook. I do not think that such efforts go far enough—accessing such data requires a lengthy process. Yet, I commend Meta's effort to at least be the first social media company to take this step (Twitter used to give researchers access to feed data, but recently stopped doing so). Clearly, Meta's action does not stem from its commitment to transparency. Yet, the step represents a signal that pressure from the public (and, hopefully, from book such as this one) has become consequential. And actors who step in to share data due to laws before such laws force them to will look better than others. Again, the current hypercompetitive environment leads companies to take transparency-washing actions, which, in my opinion, is better than nothing.

I hope that public pressure will also drive more diverse hiring practices and more vetted algorithms. The general public has increasingly come to learn about nudging algorithms, data leaks, and underregulated data broker markets. These data brokers often sell data to law enforcement agencies, which use it for predicting policing, a discriminating way to exercise power. I think that how academics educate future leaders (i.e., people now in college)—particularly in schools involving STEM programs—will affect how organizations and other actors will handle data and build algorithms in the future. In the 1960s, Herbert Simon and Allen Newell defined computer science in a now-famous paper. They noted that computer science scholars focus on algorithms, artifacts, and the environmental consequences that result from them. Until today, computer scientists have done an excellent job in working on the first two areas; however, they have largely overlooked the third area, which involves ethical and societal consequences (some notable exceptions include the Data & Society[1] organization, the AI Now Institute,[2] and the ACM/FAccT[3] community). College educators need to introduce ethical modules into technical classes in computer and data science programs. The AI ecosystem cannot work in a vacuum with programmers that have little to no idea about what they program for and what consequences could result from their work. While lecturers may not always find it easy to teach ethics and instill particular values and beliefs in students, they can help students recognize the impact that what they will build (algorithms and artifacts) may have on society. Future leaders will shape the power balances between companies and citizens in our society. For academic readership, we need to start doing a better job in our own backyard.

Along with incorporating ethics class into computer and data science programs, colleges need to focus on recruitment and admission. The US Supreme Court recently voted to curb affirmative action in higher education.[4] Thus, even if academics (like me) commit to diversity (and do not fear saying so due to tenure privileges), administrators ultimately manage admissions. These people (called "staff," at least in the US system) do not benefit from lifetime contracts (i.e., tenure) and need to report to university leaders with the mandate to maintain the bottom line. I am privileged to work at a university with a mission that focuses on inclusion, but that does not represent the norm. Higher education remains in distress and continues to experience shock from the COVID-19 pandemic with students increasingly requiring online classes. In particular, this phenomenon affects colleges whose revenue come mainly from on-campus students (e.g., via dorms).

As a last note, I purposely (yet unfairly) neglect one topic in this book: the role algorithms and AI play in the global supply chain and the associated ethical and DEI implications. I purposely avoided a chapter on this topic because I think it deserves an entire book (maybe my next project!). But the pandemic emphasized the relevance of historical data for algorithms and AI to work in a predictive fashion. In my opinion, such

[1] https://datasociety.net.

[2] https://ainowinstitute.org.

[3] https://facctconference.org.

[4] https://www.supremecourt.gov/opinions/22pdf/20-1199_hgdj.pdf.

dependence to data clearly illustrates the limits that automated systems face. Black swan events (unpredictable events with potentially severe consequences) can crash systems and stimulate human creativity. Supply chain resilience, an emerging topic that I discussed recently in a podcast (see Spotify[5]) with an Accenture data science director, might become a new global hype. Designing automated systems that can work with just real-time data when black swan events occur will become paramount to ensure the fair and even distribution of goods and services to populations worldwide. Not focusing on resilience, will expose countries to supply disruptions, which—not surprisingly—will affect people unevenly and will penalize people at the margin.

[5] https://podcasters.spotify.com/pod/show/bentley-university-mccall/episodes/Lessons-learned-from-the-supply-chain-disruption-during-the-COVID-pandemic-e25keaq.

Appendix 1: Major US Court Rulings Discussed in the Book

Plessy v. Ferguson[6] *(1896)*

This ruling, origin of most discriminations in the 1900s, asserts that racial segregation laws do not violate the US Constitution as long as the facilities for each race were equal in quality, a doctrine that came to be known as "separate but equal." This landmark decision allowed states to reestablish racial segregation practices that were extremely common at the end of the reconstitution era (1977), especially in the Southern states, echoing the Jim Crow Laws.

Shelley v. Kraemer[7] *(1948)*

With this ruling, the Supreme Court held that racially restrictive housing covenants cannot legally be enforced. Doing otherwise would violate the Fourteenth Amendment's Equal Protection Clause, which prohibits racially restrictive housing covenants from being enforced. This ruling was hardly enforced until the 1970s.

[6] https://www.archives.gov/milestone-documents/plessy-V-ferguson.

[7] https://supreme.justia.com/cases/federal/us/334/1/.

M. Marabelli, *AI, Ethics, and Discrimination in Business*, Palgrave Studies in Equity, Diversity, Inclusion, and Indigenization in Business,
https://doi.org/10.1007/978-3-031-53919-0

Brown v. Board of Education of Topeka[8] *(1954)*

Probably one of the most impactful landmark decisions of the Supreme Court, it holds that racial segregation in public schools are unconstitutional, even if the segregated schools are otherwise equal in quality. This ruling terminates, de jure, the doctrine known as "separate but equal," supported by the previous Plessy v. Ferguson ruling. The Court ruled that "separate educational facilities are inherently unequal." Laws that impose them would violate the Fourteenth Amendments Equal Protection Clause.

Jones v. Mayer[9] *(1968)*

It enforces the Thirteenth Amendment and holds that the US Congress regulates the sales of private property to prevent racial discriminations. While racial discrimination is still a problem in housing, this ruling helped to strengthen fair housing laws.

Roe v. Wade[10] *(1973)*

This ruling, drawing on the Fourteenth Amendment's "right to privacy" (Due Process Clause) protects women's right to have an abortion. With this Supreme Court decision, states could not enforce anti-abortion laws.

Planned Parenthood of Southeastern Pa. v. Casey[11] *(1992)*

With this ruling, the Supreme Court upheld the right to have an abortion as established by the "essential holding" of Roe v. Wade and issued as its "key judgment" the restoration of the undue burden standard when evaluating state-imposed restrictions on that right. The Court overturned Roe's strict scrutiny standard of review of a state's abortion restrictions with the undue burden standard, under which abortion restrictions would be unconstitutional when they were enacted for purposely placing

[8] https://supreme.justia.com/cases/federal/us/347/483/.

[9] https://supreme.justia.com/cases/federal/us/392/409/.

[10] https://supreme.justia.com/cases/federal/us/410/113/.

[11] https://supreme.justia.com/cases/federal/us/505/833/.

a substantial obstacle in the path of a woman seeking an abortion of a nonviable fetus.

Dobbs v. Jackson Women's Health Organization[12] *(2022)*

With this ruling, the Supreme Court reverses Roe v. Wade and Planned Parenthood v. Casey. The court held that the Constitution of the United States does not confer a right to abortion. With this ruling, states can enforce abortion laws as the right of abortion is no longer a constitutional right.

Gonzalez v. Google[13] *(2023) and Twitter, Inc. v. Taamneh*[14] *(2023)*

These two cases questions whether social media platforms could be held responsible when recommendation systems are covered by liability exemption under Section 230 of the 1934 Communication Act (and the following Telecommunication Act of 1996, which concerns Internet service providers) when terrorism-related contents are posted by users and hosted on their servers. The Gonzalez v. Google case was granted *certiorari* alongside another Section 230 and terrorism-related case, Twitter, Inc. v. Taamneh. In the Twitter, Inc. v. Taamneh case, the Supreme Court ruled that the charges against the social media company were not permissible under antiterrorism laws. Gonzalez was sent back to lower courts on a *per curiam* decision with instructions to consider the Court's decision in Twitter.

[12] https://www.supremecourt.gov/opinions/21pdf/19-1392_6j37.pdf.

[13] https://www.supremecourt.gov/opinions/22pdf/21-1333_6j7a.pdf.

[14] https://www.supremecourt.gov/opinions/22pdf/21-1496_d18f.pdf.

Appendix 2: Other Relevant Laws

Bill of Rights, US (1791)[15]

Written by President and Founding Father James Madison in 1789 and ratified in 1791, it includes the first ten amendments to the US Constitution. These amendments are aimed at giving more right to American citizens, such as the right to free speech, citizenship by birth, equal right before the law.

Civil War Amendments, US (1865–1870)[16]

Also known as Reconstitution Amendments, they are the Thirteenth, Fourteenth, and Fifteenth amendments to the US constitution. They were created in the period 1865–1870 in the aftermath of the Civil War. These amendments aimed at protecting Black Americans from discriminations stemming from slavery past.

[15] https://www.archives.gov/milestone-documents/bill-of-rights.

[16] https://constitution.congress.gov/browse/essay/intro.3-4/ALDE_00000388/.

M. Marabelli, *AI, Ethics, and Discrimination in Business*, Palgrave Studies in Equity, Diversity, Inclusion, and Indigenization in Business,
https://doi.org/10.1007/978-3-031-53919-0

Jim Crow Laws (Circa 1880)[17]

Introduced and enforced in the Southern States of the US during the period 1880–1920, these laws established racial segregation in schools, neighborhoods and the workplace and remained in force in some states until the 1964 Civil Rights Act.

Civil Rights Act (1964)[18]

It bans discriminations based on race, color, religion, sex, and country of origin and, in particular, prohibits discriminations in schools, in the workplace, and during voting procedures. The act was proposed by President Kennedy but eventually passed under the Johnson administration.

Fair Housing Act, US (1968)[19]

Signed into law by President Johnson, it is actually part of the Civil Rights Act of 1968 (in the aftermath of the assassination of Martin Luther King). Titles VIII and IX of the Civil Rights Act (known as the Fair Housing Act). It expands on the 1866 Civil Rights Act, and prohibited discrimination concerning the sale, rental, and financing of real estates based on race, religion, country of origin and, since 1974, sex, since 1988 people with disabilities and families with children.

GDPR, EU (2018)[20]

The General Data Protection Regulation (GDPR), implemented in 2018, focuses on information privacy and had its most impacts on online data. It aims at giving EU citizens control over their data and is often used to determine the extent to which Internet platforms (social media, e-commerce websites etc.) can collect and use user data. It is centered on user consent, which needs to be explicit.

[17] https://www.loc.gov/classroom-materials/jim-crow-segregation/.

[18] https://www.archives.gov/milestone-documents/civil-rights-act.

[19] https://www.justice.gov/crt/fair-housing-act-1.

[20] https://gdpr.eu.

EU AI Act (2021)[21]

It is a proposed EU regulation aimed at generating a comprehensive framework disciplining how artificial intelligence (AI) systems are used in society (with the exception of the military). It classifies AI based on the potential harm it can cause to individuals. The classification primarily falls into three categories: banned practices, high-risk systems, and other AI systems.

PIPL, China (2021)[22]

The Personal Information Protection Law (PIPL) applies to China's private companies (even if it might apply to companies operating abroad) and aims to protect citizens' data, in a way that it similar to GDPR (even if GDPR applies also to government institutions). It also regulates the transfer of personal data outside China.

[21] https://www.europarl.europa.eu/news/en/headlines/society/20230601STO93804/eu-ai-act-first-regulation-on-artificial-intelligence.

[22] http://en.npc.gov.cn.cdurl.cn/2021-12/29/c_694559.htm.

Index

M. Marabelli, *AI, Ethics, and Discrimination in Business*, Palgrave Studies in Equity, Diversity, Inclusion, and Indigenization in Business,
https://doi.org/10.1007/978-3-031-53919-0

GPSR Compliance
The European Union's (EU) General Product Safety Regulation (GPSR) is a set of rules that requires consumer products to be safe and our obligations to ensure this.

If you have any concerns about our products, you can contact us on

ProductSafety@springernature.com

In case Publisher is established outside the EU, the EU authorized representative is:

Springer Nature Customer Service Center GmbH
Europaplatz 3
69115 Heidelberg, Germany

www.ingramcontent.com/pod-product-compliance
Ingram Content Group UK Ltd.
Pitfield, Milton Keynes, MK11 3LW, UK
UKHW022222051026
15624UKWH00002B/294
9783031539183